W0050886

KEY TOPICS
IN BRAIN
RESEARCH

Edited by A. Carlsson, P. Riederer,
H. Beckmann, T. Nagatsu,
and S. Gershon

P. Riederer and M. B. H. Youdim (eds.)

Iron in Central Nervous System Disorders

Springer-Verlag Wien New York

Prof. Dr. Peter Riederer
Department of Psychiatry, University of Würzburg,
Federal Republic of Germany

Prof. Dr. M. B. H. Youdim
Department of Pharmacology, Faculty of Medicine,
Technion, Haifa, Israel

This work is subject to copyright.
All rights are reserved, whether the whole or part of the material is concerned,
specifically those of translation, reprinting, re-use of illustrations, broadcasting,
reproduction by photocopying machines or similar means, and storage in data
banks.

© 1993 Springer-Verlag/Wien

Product Liability: The publisher can give no guarantee for information about
drug dosage and application thereof contained in this book. In every individual
case the respective user must check its accuracy by consulting other pharmaceutical literature. The use of registered names, trademarks, etc. in this publication
does not imply, even in the absence of a specific statement, that such names are
exempt from the relevant protective laws and regulations and therefore free for
general use.

Printing: Eugen Ketterl GesmbH, A-1180 Wien
Printed on acid-free and chlorine-free bleached paper

With 50 Figures

ISSN 0934-1420

ISBN-13: 978-3-211-82520-4   e-ISBN-13: 978-3-7091-9322-8
DOI: 10.1007/978-3-7091-9322-8

Preface

The role of the metals copper, zinc, magnesium, lead, manganese, mercury, lithium and aluminium in neuropsychiatric disease are well known and has been discussed on several occasions. Yet little attention has been paid to iron, the most abundant transitional metal in the body and the earth's crust. Iron plays a major role as a cofactor of numerous metabolic enzymes, it is important for DNA and protein synthesis, and has a crucial role in the oxygen carrying capacity of haemoglobin.

Some of the most devastating diseases of systemic organs are associated with abnormal iron metabolism. Yet only very recently its role in the central nervous system has been considered. Thus nutritional iron deficiency and iron overload afflict some 500–600 million people. It is also well recognized that too little or too much iron can produce profound effects on the metabolic state of the cell, and therefore the regulation of iron uptake and disposition is tightly relegated by the cell. Its transport into the cell and storage are handled by transferrin, ferritin and haemosiderin. Nowhere are these processes so well recognized as in the case of brain iron metabolism. Iron does not have ready access to the adult brain as it does to other tissues, since it does not cross the blood brain barrier (BBB). All the iron present in brain is deposited before the closure of BBB at an early age where it is sequestered and conserved. Therefore its turnover is extremely slow. Nevertheless nutritional iron deficiency can produce loss of brain iron and induce biochemical and behavioural changes. This can become irreversible if the deficiency occurs early in life and is for a prolonged period. By contrast, the presence of excess iron in specific brain regions, which is the subject of the present volume, may be associated with the primary or secondary cause of neurodegeneration, and neurological diseases such as Parkinson's disease, Alzheimer's disease, amyotrophic lateral sclerosis, Hallervorden-Spatz disease, and stroke and trauma. The main feature of iron neurotoxicity, which is dealt with at length, is its role in the production of oxygen free radicals and their participation in oxidative stress initiated neurodegeneration. The ability to chelate iron or to scavenge oxygen free radicals as a means of treating neurological diseases or of neuroprotection are considered, appreciating the problems of drug transport across the BBB. To investigate these

systems a number of animal models and neuronal cell culture preparations have been employed. These studies have clearly demonstrated that
iron can produce neuronal degeneration in vivo and in cell culture studies
and that these can be prevented by iron chelators and radical scavengers.
However, the role of iron in neurological disease remains elusive. Nevertheless we believe that the content of this volume will illuminate the
importance for normal brain function of maintaining iron homeostasis
and that the therapeutic approaches to some, if not all, neurological
diseases may involve scavenging of oxygen free radicals.

December 1993   P. RIEDERER
                                                  M. B. H. YOUDIM

Contents

Cellular and regional maintenance of iron homeostasis in the brain: normal and diseased states

J. R. Connor

George M. Leader Family Laboratory for Alzheimer's Disease Research, Department of Neuroscience and Anatomy, Pennsylvania State University, and M. S. Hershey Medical Center, Hershey, Pennsylvania, U.S.A.

Summary

Iron is an essential trophic factor for normal development of the brain and for maintenance of normal neurological function throughout life. However, iron when not regulated, can become a potent toxin through its ability to induce lipid peroxidative damage. Consequently, the brain has an exquisite system to assure the availability and timely delivery of iron. In this manuscript we review the cellular and regional distribution of the proteins involved in mediating the regulation of iron in the brain. These proteins, transferrin and its receptor (iron mobilization) and ferritin (iron storage) are thus responsible (along with the cells in which they reside) for maintaining iron homeostasis in the brain. Within the brain, iron and iron regulatory proteins have a region specific distribution being especially abundant in areas associated with motor functions. At the cellular level oligodendrocytes are the predominant cell type to stain for iron, transferrin and ferritin; although, numerous ferritin-positive microglial cells are present. Preliminary data from an analysis of isoforms of ferritin in the brain which may provide insight into the role of each cell type in iron regulation are discussed herein.

We further review the cellular and regional alterations which occur in the brain in such diseases as Alzheimer's, Parkinson's and Multiple Sclerosis in which a disruption in iron homeostasis may be part of the pathogenesis of these diseases.

A high iron requirement for normal brain function both during normal development and in adulthood has been established. During development, iron deficiency is associated with cognitive and motor impairment which persist into adulthood (Yehuda and Youdim, 1988). Some evidence exists that the mental and developmental retardation associated with in utero exposure to ethanol may be mediated through disruptions in brain iron metabolism (Dicker et al., 1990; Nordmann et al., 1990; Reinke et al., 1990).

In the adult, severe, chronic iron deficiency must occur before any neurological deficits are realized (Yehuda and Youdim, 1988). This observation suggests that the brain has a considerable storage capacity for iron which is readily mobilizable. Although, the amount of iron in the brain varies with each brain region, the basal ganglia have among the highest levels of iron in the brain; reportedly as much iron per unit weight as the liver (Hallgren and Sourander, 1958). Brain iron reportedly has a very slow turnover rate (Dallman and Spirito, 1977). The levels of iron in each brain region however are not static and have been shown to increase with age (Hallgren and Sourander, 1985; reviewed in Connor, 1992) and some disease states are associated with excessive accumulation of iron. In Alzheimer's disease, specific brain regions such as the hippocampus, nucleus basalis of Meynert, and the superior temporal gyrus are all known to contain elevated iron relative to normal (Thompson et al., 1988; Connor et al., 1992). Increased iron levels are also reported in Parkinson's disease in brain regions which are specifically affected in this disease (Sofic et al., 1988). Recently, magnetic resonance imaging (MRI) techniques have been used to reveal an increased accumulation of iron in the basal ganglia in multiple sclerosis (Drayer et al., 1986). In each of these diseases, oxidative damage to proteins is considered a key component of the pathogenesis and the role of iron in the induction of oxidative stress in the brain is well established (see other chapters in this publication).

At the cellular level, the histological examination of iron was greatly enhanced by a modification of the Perl's reaction which intensified the iron reaction product with 3'3 diaminobenzidine (Nguyen-Legros et al., 1980). Through the use of this modification, histological analyses of iron distribution in brain are now more closely related to the biochemical data. For example, biochemical analysis revealed relatively high iron levels in white matter of the brain (Rajan et al., 1976; Ehmann et al., 1986), but histochemically iron was not detectable (Diezel, 1955). However, it is now clear that iron is abundant in white matter (Connor and Menzies, 1990; Connor et al., 1990; Morris et al., 1992) and occurs in patches of cells which stain intensely for iron. Another modification for maximizing iron staining in the brain (and probably in other tissue) is in regard to tissue processing. The tissue should be fixed in 10% neutral *buffered* formalin and processed for staining within 2 weeks of fixation. Non-paraffin processed tissue (vibratome or sucrose treated frozen sections) yields superior results in all cases.

The predominant cell type to contain iron in human brain tissue (Connor et al., 1990; Morris et al., 1992; Dwork, 1988), monkey

(Francois et al., 1981), rat (Hill and Switzer, 1984; Connor and Menzies, 1990) and mouse (Levine and Macklin, 1990) are oligodendrocytes. Even in iron rich areas such as the caudate-putamen, substantia nigra and deep cerebellar nuclei it is the oligodendrocytes which stain robustly for iron (Benkovic and Connor, 1993). Fibers in the neuropil stain intensely in the deep cerebellar nuclei and substantia nigra; fibrous staining in the basal ganglia is confined to the white matter tracts which form the striations (Fig. 1A; Connor et al., 1990). Neurons, particularly pyramidal neurons in the cerebral cortex and hippocampus (as well as granule cells in the latter region) have small punctatum of iron reaction product in their somata which increase in density in rats with age (Benkovic and Connor, 1993). Another cell type which stains prominently for iron are tanycytes which line the third ventricle (Hill and Switzer, 1984; Benkovic and Connor, 1993) these cells may be involved in transporting iron between the brain and cerebrospinal fluid. This avenue of iron entry into the brain has not to our knowledge been explored.

Disruptions in the histological distribution of iron have been reported in some disease states. In Alzheimer's disease (AD), iron staining is associated with neuritic plaques and iron encrustation along blood vessels in AD has also been reported (Connor et al., 1991; Goodman, 1953). Iron is also found in plaques in patients who had multiple sclerosis (1982). Neuromelanin-containing neurons of the substantia nigra accumulate iron in Parkinson's disease (Good et al., 1992). If the population of oligodendrocytes is compromised as in myelin mutants, iron is found in astrocytes and microglial cells rather than the dysfunctional oligodendrocytes (Connor and Menzies, 1990). Siderotic microglia are also present in autopsy material from HIV infected individuals (Gelman et al., 1992) and in an animal model of superficial siderosis (Koeppen and Dentinger, 1988). These latter observations indicate that transport of iron into the brain continues in pathological states, but the uptake of iron into cells in the brain is completely altered. Whether the uptake of iron into astrocytes and microglia is for sequestration and detoxification or whether the iron is required as part of the response of these cells to brain insult is not known. Another unknown, is the factors which regulate cellular iron uptake into each of the distinct cell populations. Continued iron uptake into the brain and cellular redistribution indicates that relying on quantitative data alone when assessing whether metal (or protein) function is normal may provide a false sense of homeostasis.

Maintenance of iron homeostasis in the brain both cellularly and regionally likely involves the same proteins which are involved systemically in iron transport and sequestration. Those proteins are transferrin and its receptor and ferritin which have all been found in the brain and are discussed below.

Transferrin

Transferrin (Tf) is the major iron transport protein in the body. In addition to iron transport, Tf also transports manganese (Aschner and Aschner, 1990) and aluminum (Roskams and Connor, 1990) into the brain. Consequently, Tf may play a critical role in the transport of metals other than iron in the brain and may thus be important to studies on general metal neurotoxicity.

Transferrin mRNA has been detected in the brain within oligodendrocytes (Bloch et al., 1985) and the choroid plexus (Aldred et al., 1987). Indeed, in this latter organ Tf mRNA levels are similar to those in liver on a per unit weight basis. The choroid plexus epithelial cells secrete Tf (Tsutsumi and Sanders-Bush, 1990) which is relatively high in the cerebrospinal fluid (Elovaara et al., 1985) suggesting the ventricular system may have a significant role in the maintenance of iron homeostasis in the brain.

The brain is the only organ in which a postnatal increase in Tf mRNA occurs (Levine et al., 1984). In animals in which the oligodendrocytic population fails to thrive, Tf mRNA levels do not increase postnatally beyond those seen at postnatal day 5 (Bartlett et al., 1991). This observation strongly suggests that a mature oligodendrocyte population is required for normal Tf mRNA expression, and is consistent with the in situ studies that Tf mRNA is found only in oligodendrocytes. The dependence on Tf mRNA expression on oligodendrocytes raised the question whether the expression of Tf mRNA is also dependent on myelin expression. Investigation of shiverer mice myelin mutants in which oligodendrocytes are numerically normal but do not synthesize normal myelin (qualitatively or quantitatively) revealed normal Tf mRNA levels in brain (unpublished observations). These data indicate Tf mRNA in brain is more closely associated with oligodendrocytes than myelin production. Furthermore, because Tf protein expression is diminished in shiverer mouse brains, these data imply Tf expression may be regulated at the level of translation. Factors which regulate Tf expression in brain and myelin expression in brain are unknown and one aspect of our research is dedicated to exploring relationships between myelin production and iron metabolism in oligodendrocytes.

A temporal relationship between Tf/iron and myelination in the postnatal brain has been established where Tf (and iron) uptake into the brain is highest at the time of peak onset of myelinogenesis (Taylor and Morgan, 1990). A report that hypomyelination is associated with iron deficiency (Larkin and Rao, 1990) in rodents further supports a role for Tf/iron in myelin production and maintenance. Postnatal requirements for iron (as mediated by Tf) have been further demonstrated in an animal (hypotransferrinemic mice; Huggenvik et al., 1989) in which endogenous Tf is not made due to a splicing defect in the Tf gene, histological abnormalities are present in adult animals in regions in which a postnatal development is significant even though the animal has received systemic injections of Tf over its life-time (unpublished observations).

Immunohistochemical studies on Tf in the brain reveal that oligodendrocytes are the predominant cell containing Tf (Fig. 1 B; Connor et al., 1990; Connor and Fine, 1986) which is consistent with the in situ analysis for Tf mRNA. Some neuronal staining for Tf has been reported during development (Oh et al., 1986) and in adult human brain material (Connor et al., 1990; Dwork et al., 1988) but these results are inconsistent and difficult to interpret. Injections of colchicine to block intracellular transport of Tf did not affect neuronal staining with Tf (Connor and Fine, 1986). Under conditions in which the oligodendrocyte population fails to thrive, the levels of Tf in the brain are < 5% of normal (Connor et al., 1987) supporting the immunohistochemical studies that oligodendrocytes are the predominant cell type to contain Tf.

Altered cellular distribution of Tf has been reported. Astrocytes in white matter in AD brain tissue contain Tf (Connor et al., 1991) as do those surrounding plaques in multiple sclerosis (Esiri et al., 1976) and in

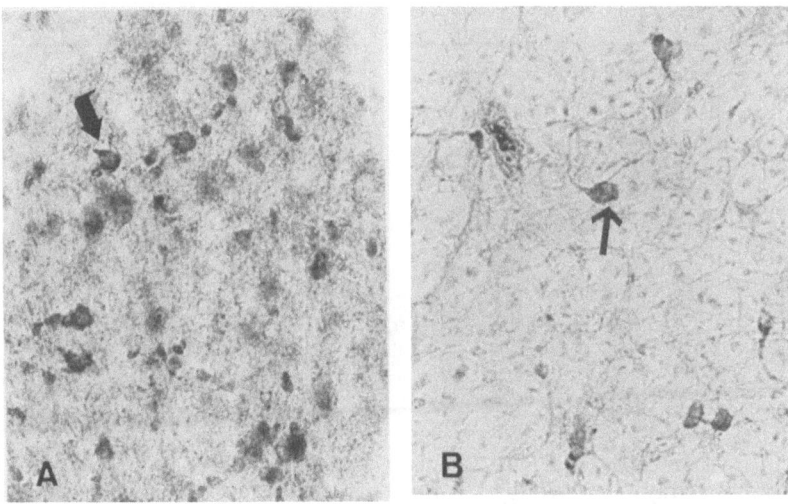

Fig. 1. A Iron staining (Perl's reaction plus intensification) in the human corpus striatrum. The fibers in the background are stained. The stained cells are found within the striated area as well as in the gray matter and resemble oligodendrocytes (e.g. arrow). 600×. B Transferrin immunostaining in the human striatum is confined mostly to oligodendrocytes in the white matter (e.g. arrow). 600×

central pontine myelinolysis (Gocht and Lohler, 1990). Factors which may stimulate Tf uptake and/or expression (Espinosa et al., 1990) within astrocytes are currently unknown. Elucidation of these factors could be significant to studies concerned with understanding ways in which the brain may protect itself from iron-induced oxidative damage.

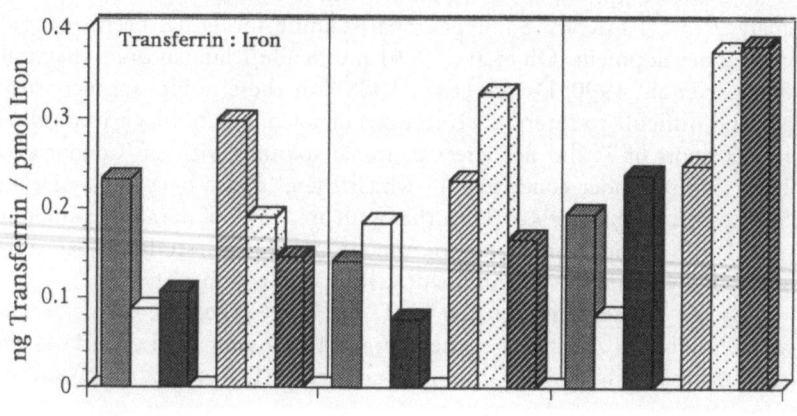

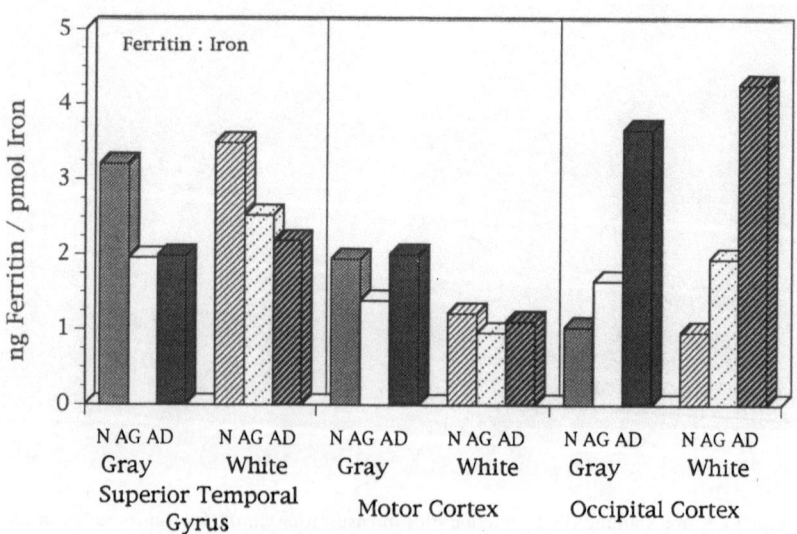

Fig. 2. Bar Graph: The ratio of transferrin to iron (top) and ferritin to iron (bottom) in three regions of the human cerebral cortex. Details on the tissue samples are presented in Connor et al. (1992) (*N* normal adults, *AG* normal aged controls, *AD* Alzheimer's disease)

Quantitative analysis of Tf has been performed on rat, mouse and human brain tissue. In the adult rat (75 days), Tf levels range from 4–6 μg/mg protein (unpublished observations). The highest levels (6 μg/mg) are in the midbrain whereas the cerebellum-pons and cerebral cortex are lower (4 μg/mg). During development (17 days of age) the pons-cerebellum contain more Tf per unit wet weight and per mg protein than the cerebral cortex (Connor et al., 1987). In the mouse, Tf levels are slightly lower (2–4 μg/mg protein) than in the rat throughout the brain (unpublished observations).

In human brain tissue (Connor et al., 1992), Tf levels in cerebral cortical regions range between 1.0 and 8.0 μg/mg protein; and are highest in the motor cortex (as is ferritin and iron). The levels of Tf in the subcortical white matter are higher than those in the corresponding gray matter. With age the ratio of transferrin to iron (Fig. 2) decreases by over 50% in some brain regions such as the superior temporal gyrus which undergoes significant atrophy with normal aging and in AD (Perry, 1986). In the motor cortex, where considerable histopathological abnormalities occur in AD, the brain Tf:iron ratio increases with normal aging but decreases dramatically in AD. The occipital cortex which shows little change with normal aging or in AD and shows an inexplicable decrease in the Tf:iron ratio with normal aging in the gray matter but no change in AD and an increase in the ratio in white matter in both aging and AD.

The tendency for Tf: iron ratio to decrease with age and to a greater extent in AD tissue has lead to our hypothesis of a decrease in iron mobility in AD. A loss of iron mobility would be manifested in decreases in oxidative metabolism and increased susceptibility to iron-induced oxidative damage which are both established phenomena in the AD brain. The ratio data in relationship to our decreased iron mobility hypothesis may be more significant in the white matter than in the gray because in white matter the cellular distribution for Tf is abnormal; astrocytic immunostaining for Tf predominates over oligodendrocyte staining (Connor et al., 1991).

Ferritin

Ferritin accounts for 1/3 to 3/4 of the brain iron (Hallgren and Sourander, 1958; Octave et al., 1983), but is the least studied of the major components of the iron regulatory system in the brain. Like transferrin, ferritin may also bind metals other than iron and thus investigation of this protein may also be relevant to general metal neurotoxicity.

Ferritin is found within microglia (Kaneko et al., 1989; Koeppen and Dentinger, 1988; Connor et al., 1990) and oligodendrocytes (Connor et al., 1990) in human and rat (Benkovic and Connor, 1993) brain tissue. Ferritin positive astrocytes have been observed but are mostly confined to the basal ganglia where they have been noted to increase with age (Connor et al., 1990). Ferritin immunoreactivity has been observed in cells which are associated with neuritic plaques in AD (Connor et al., 1991; Kaneko et al., 1989). Quantitatively, ferritin ranges between 10 and 45 µg/mg protein in the human cerebral cortex, thus it is considerably more abundant than transferrin (Connor et al., 1992). Brain ferritin levels are both region specific and fluctuate within gray and white matter. In the occipital and temporal cortices, ferritin is higher in the white matter than in the gray whereas, in the motor cortex, ferritin is higher in the gray matter relative to the underlying white matter (Connor et al., 1992). Ferritin is higher in extrapyramidal regions such as the globus pallidus, substantia nigra and caudate nucleus than in the cerebral cortex (Broadmen area 10) on a per unit wet weight basis (Dexter et al., 1990, 1991).

Ferritin levels in the human brain have been examined in AD and Parkinson's disease. In the cerebral cortex, ferritin decreases with normal aging and the decrease is exacerbated in AD except in the occipital cortex where it increases in both the aged and AD groups (Connor et al., 1992). The ratio of ferritin to iron decreases by over 1/3 in the superior temporal gyrus with both age and AD whereas there is little change in the ratios in the motor cortex and an actual increase (3X) in the occipital cortex (Fig. 2). These data would suggest ferritin in the temporal region contains more iron per mol which has been suggested (Fleming and Joshi, 1987) whereas in the occipital cortex there is less iron per mol of ferritin with age. Whether the increase in ferritin without an increase in iron could serve as a protective mechanism is not known at present. It is interesting that the cortical region which is the least susceptible to AD related atrophy shows the highest ferritin:iron ratio whereas the region most susceptible to both age and AD has a decrease in the ferritin : iron ratio. Any significance placed on the altered ratios of both Tf : iron and ferritin : iron as having a causative role in the enhanced susceptibility of the superior temporal gyrus to neurodegeneration in AD at this time would be speculative.

In Parkinson's disease, ferritin has been reported to both increase in the substantia nigra (Sofic et al., 1988) and decrease (Dexter et al., 1990, 1991) relative to normal. One possible explanation given for the discrepancy in these results is the type of antibody used. Ferritin is made of two isoforms; heavy (H-chain) and light (L-chain). We have recently obtained monoclonal antibodies to the H and L isoforms of ferritin and performed

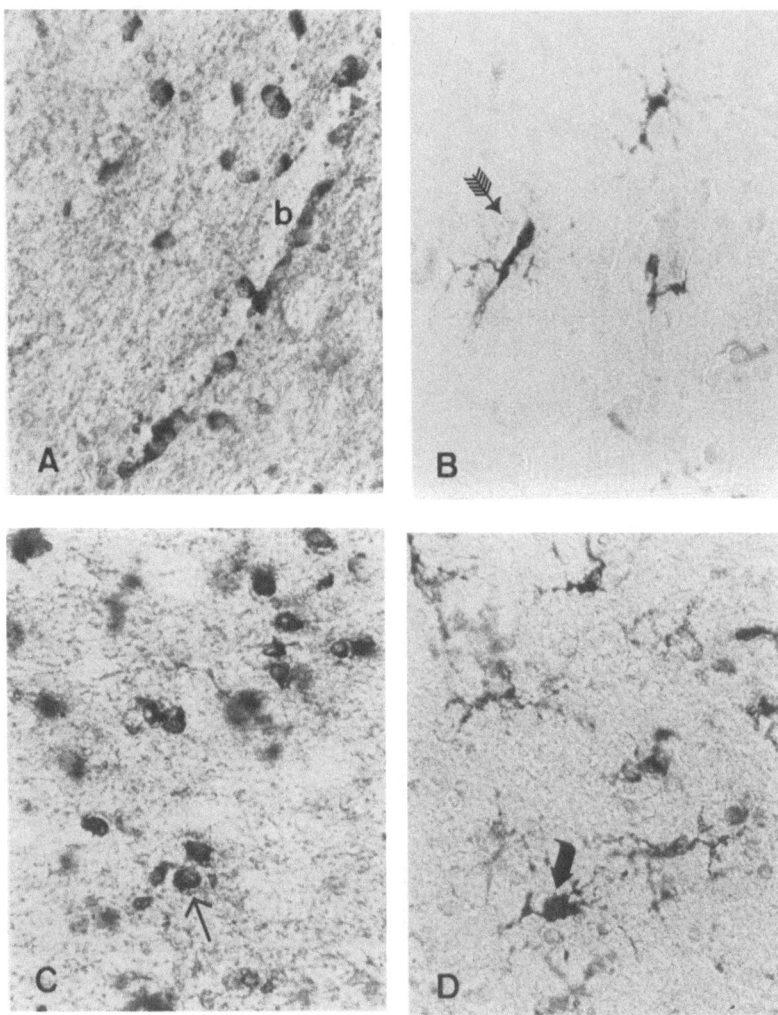

Fig. 3. A H-chain ferritin from the human temporal cortical region is found in oligo-dendrocytes. The cells are disperesed throughout the cortex in all layers and can be found in perineuronal and perivascular positions. Perivasallar cells can be seen in this micrograph associated with a blood vessel (*b*) which is coursing through the section. 600×. B L-chain ferritin is dramatically different from H-chain in the human cerebral cortex. Microglial cells (e.g. arrow) predominantly immunostain in the cerebral cortex with the L isoform. Non-stained neurons can be seen in the background. 600×. C In human subcortical white matter, H-chain ferritin is found in oligodendrocytes (e.g. arrow). 750×. D Microglial cells in human subcortical white matter are immunoreactive with the L-chain ferritin isoform. These cells are easily distinguished by their amorphous soma and short, occasionally heavy processes (e.g. arrow)

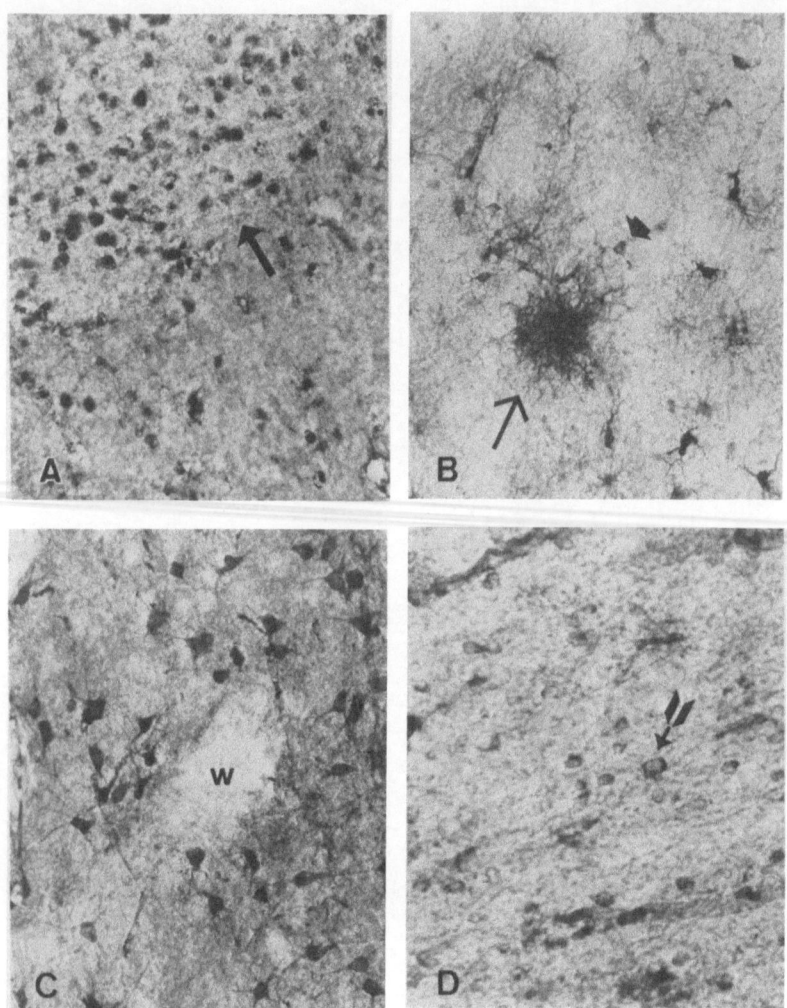

Fig. 4. A In the human striatum, H-chain ferritin labels oligodendrocytes in both the gray and white matter (arrow demarks gray-white boundary), but the fibers in the white matter are not stained. 300×. B L-chain ferritin in the human striatum labels astrocytes (arrow) as well as microglial cells (arrowhead). 300×. C Transferrin receptor labelling in the human striatum is intense and localized to neurons. The white matter striations (w) are unlabelled. 300×. D In the human subcortical white matter, transferrin receptor labelling is light and associated with oligodendrocytes (e.g. arrow). 400×

immunohistochemistry on monkey and human brain tissue. In the monkey, neurons immunostain for only the H-chain isoform whereas oligodendrocytes immunostain for both H and L ferritin. Microglial cells robustly label with L-chain ferritin, but not H-chain (Connor et al., 1992). However, in human brain tissue, neurons do not immunostain for H or L ferritin consistent with existing reports that ferritin is not present in neurons. The rest of the staining pattern for H and L ferritins in the human brain is similar to that seen in the monkey. Oligodendrocytes immunostain both H and L chain ferritin in the gray matter (Fig. 3 A) and white matter (Fig. 3 C). Microglial cells however, intensely label only with L chain in both gray (Fig. 3 B) and white matter (Fig. 3 D). Astrocytes do not normally contain either H or L chain ferritin except in the caudate where they are intensely and specifically reactive for the L isoform of ferritin (Fig. 4 B compare to Fig. 4 A) . This latter observation is consistent with our earlier report on the human tissue (Connor et al., 1990).

The functional distinction between the two isoforms of ferritin are not established, but some differences are being investigated. The H-chain ferritin is generally associated with organs in which iron is more rapidly utilized whereas the L-isoform is more closely associated with longer storage of iron (Arosio et al., 1991). The significance of the distinct cellular distributions lies in the determination of the functional distinctions between the 2 isoforms and as the research develops should shed considerable light on the role of each of the cell types in iron regulation within the brain and perhaps the ability of each cell type to withstand oxidative damage.

Transferrin receptor

Iron uptake into the brain is a continuous process throughout life and involves utilization of a transferrin receptor located on endothelial cells of brain microvasculature (Jefferies et al., 1984; Fishman et al., 1987; Pardridge et al., 1987). Tf receptors have been demonstrated in the rat brain using ^{125}I-Tf (Hill et al., 1985; Mash et al., 1990). These studies provide data on the regional distribution of the Tf receptors and are in good agreement that regions associated with motor function have the highest density of receptors. However, the autoradiographic studies do not provide sufficient resolution to distinguish between neural tissue binding and microvasculature binding. This is a significant consideration because the microvasculature in the brain has a 6–10 fold higher concentration of receptors than the parenchyma (Kalaria et al., 1992) and the vasculature is more dense in the gray matter. Thus, the receptor distribution in the autoradiographic studies could be reflecting the distribution of the blood vessels. In possible support of this idea is the lack of Tf receptor distribution in white matter despite a report of a high affinity Tf receptor on oligodendrocytes in culture (Espinosa and Focaud, 1987). Furthermore, an autoradiographic study on iron-binding sites in the rat brain using $^{59}FeCl_3$ (Barkai et al., 1991) rather than ^{125}I-Tf found considerable binding in the white matter.

A strong positive correlation between Tf receptor distribution (as determined by ^3H-Tf autoradiography) and cytochrome oxidase distribution has been reported in human brain tissue leading the authors to suggest that Tf receptors may provide a useful marker for neuronal respiratory activity (Morris et al., 1992). Such a relationship between Tf receptor density and cytochrome oxidase has also been indicated in the rat brain (Mash et al., 1990) .

Tf receptor analyses of brain homogenates have found the affinity of the brain Tf receptor for Tf is higher than in other cells or organs studied. The Kd for Tf in the brain ranges from 1–10 nM (Hill et al., 1985; Mash et al., 1990; Kalaria et al., 1992; Roskams and Connor, 1990). If Tf is conjugated with aluminum rather than iron, this affinity decreases only slightly (13 nM) which is well within physiological limits (Roskams and Connor, 1990). Consequently, aluminum could exploit the extant pathway for iron delivery to the brain and may thus enter the brain without a disruption in the blood-brain-barrier thereby negating a long-standing argument against a causative role for aluminum in neurodegenerative diseases such as AD. In this regard it is potentially significant that the nucleus basalis of Meynert has a relatively high density of Tf receptors and this area is susceptible to degeneration in AD (Morris et al., 1989).

Immunohistochemical attempts at determining the cellular distribution of Tf receptor in the brain have been hampered by the relative lack of reactivity of monoclonal antibodies to the Tf receptor in the brain (but not the brain microvasculature). A single immunoreactive neuron was shown in one of the initial reports in this field (Jefferies et al., 1984). More recently, using a Tf receptor monoclonal antibody (clone OX 26) weak immunoreaction was observed on neurons and oligodendrocytes in rat brain (Giometto et al., 1990). We have obtained a polyclonal antibody to the human placental Tf receptor (Paul Seligman; University of Colorado) which has yielded consistent results on brain tissue. Neurons in the human cerebral cortex express Tf receptors as do those in the caudate-putamen (Fig. 4 C). In subcortical white matter there is a light immunoreaction associated with oligodendrocytes (Fig. 4 D). The cellular distribution of the Tf receptor and factors which regulate receptor expression require considerable additional investigation because of reported changes which occur with age and following injury.

Following axotomy, neurons in the facial nucleus upregulate Tf receptor expression (Graber et al., 1989) indicating iron uptake into neurons may be an important component of the reparative process. Iron-binding in the rat brain is reportedly decreased with age and in the presence of free radicals which may be related observations (Barkai et al., 1991). In Alzheimer's disease, the density of Tf receptors on brain microvasculature was not changed compared to normal, but significant decreases in receptor density were noted in the hippocampus and temporal and occipital cortices. Other brain regions such as the parietal and frontal cortices and cerebellum had no change in Tf receptor density in normal versus AD brains (Kalaria et al., 1992).

During development Tf receptors are present on oligodendrocytes (especially those in the vicinity of blood vessels) in the developing optic nerve, but the detection of the receptor is lost as the animal matures. There is no expression of Tf receptors on oligodendrocytes in optic nerves from myelin deficient (md) rats in which the oligodendrocytes fail to thrive (Lin and Connor, 1989). Biochemical analysis of the Tf receptor in md rats (in which the microvasculature has been removed) reveals a 2/3 loss of Tf receptors suggesting the presence of oligodendrocytes during development is necessary for the normal expression of Tf receptors (Roskams and Connor, 1992).

Conclusion

The predominant cell in the brain in which iron and the iron regulatory proteins are found is the oligodendrocyte. The established function for oligodendrocytes is the production and maintenance of myelin. Fatty acid synthesis and cholesterol synthesis, both of which occur at a relatively high level in oligodendrocytes and are major components of myelin require iron (Larkin and Rao, 1990). Recent evidence that Tf is secreted by oligodendrocytes (Espinosa et al., 1990) suggests that Tf/iron may have a paracrine function rather than or in addition to an intracellular role. In this regard, it is tempting to speculate that the perineuronal oligodendrocytes are involved in transport of iron between neurons. The role of the other glial cells in the brain seem to be primarily that of sequestration based on the immunohistochemical staining with L-chain ferritin. Neurons on the other hand may have a high iron utilization and turnover rate based on the low level of iron staining and the presence of H-chain ferritin.

Considerable attention has recently been focused on controlling the availability of iron in the brain (see chapter by Hall) and exploiting the extant system for iron delivery for therapeutic delivery of trophic factors

(Friden et al., 1993). Although iron imbalances have been demonstrated in many neurological diseases, much basic information on iron regulation under normal conditions remains to be understood.

Acknowledgements

The work in this manuscript was supported by grants from the United States Public Health Service (NS22671, AG09063) and from the National Multiple Sclerosis Society (RG2118). The author gratefully acknowledges the assistance of Sharon Menzies and Brian Snyder in the preparation of the figures and data included in this manuscript.

References

Aldred AR, Dickson FW, Marley PD, Schreiber G (1987) Distribution of transferrin synthesis in brain and other tissues in the rat. J Biol Chem 262: 5293–5297

Arosio P, Levi S, Santambrogio P, Cozzi A, Luzzago A, Cesareni G, Albertini A (1991) Structural and functional studies of human ferritin H and L chains. In: Albertini A, Lenfant CL, Mannucci PM, Sixma JJ (eds) Biotechnology of plasma proteins. Karger, Basel (Curr Stud Hematol Blood Trans: 127–131)

Aschner M, Aschner JL (1990), Mangabese transport across the blood brain barrier: relationship to iron homeostasis. Brain Res Bull 24: 857

Barkai AI, Durkin M, Dwork AJ, Nelson HD (1991) Autoradiographic study of iron-binding sites in the rat brain: distribution and relationship to aging. J Neurosci Res 29: 390–395

Bartlett WP, Li X-S, Connor JR (1991) Expression of transferrin mRNA in the CNS of normal and jimpy mice. J Neurochem 57: 318

Benkovic S, Connor JR (1993) Ferritin, transferrin and iron in normal and aged rat brains. J Comp Neurol 337: 1

Bloch B, Popovici T, Lovin MJ, Tuil D, Rahn A (1985) Transferrin gene expression visualized in oligodendrocytes of the rat brain using in situ hybridization and immunohistochemistry. Proc Natl Acad Sci USA 82: 6706

Connor JR (1992) Proteins of iron regulation in the brain in Alzheimer's disease. In: Lauffer RB (ed) Iron and human disease. CRC Press, Boca Raton, pp 365–393

Connor JR, Fine RE (1986) The distribution of transferrin immunoreactivity in the rat central nervous system. Brain Res 368: 319–328

Connor JR, Menzies SL (1990) Altered cellular distribution of iron in the central nervous system of myelin deficient rats. Neuroscience 34: 265

Connor JR, Phillips TM, Lakshman MR, Barron KD, Fine RE, Csiza CK (1987) Regional variation in the levels of transferrin in the CNS of normal and myelin-deficient rats. J Neurochem 49: 1523–1529

Connor JR, Menzies SL, St. Martin S, Mufson EJ (1990) The cellular distribution of transferrin, ferritin and iron in the human brain. J Neurosci Res 27: 595

Connor JR, Menzies SL, St. Martin S, Fine RE, Mufson EJ (1992) Altered cellular distribution of transferrin, ferritin and iron Alzheimer's disease brains. J Neurosci Res 31: 75–83

Connor JR, Boeshore KL, Benkovic SA (1992) Isoforms of ferritin have a distinct cellular distribution in the brain. Mol Biol Cell 3: 84A

Connor JR, Snyder BS, Beard JL, Fine RE, Mufson EJ (1992) The regional distribution of iron and iron regulatory proteins in the brain in aging and Alzheimer's disease. J Neurosci Res 31: 327–335

Craelius W, Migdal MW, Luessenhop CP, Sugar A, Mihalakis I (1982) Iron deposits surrounding multiple sclerosis plaques. Arch Pathol Lab Med 106: 397

Dallman PR, Spirito RA (1977) Brain iron in the rat: extremely slow turnover in normal rat may explain the long-lasting effect of early iron-deficiency. J Nutr 107: 1075–1081

Dexter D, Carayon A, Vidailhet M, Ruberg M, Agid F, Agid Y, Lees AJ, Wells FR, Jenner P, Marsden CD (1990) Decreased ferritin levels in brain in Parkinson's disease. J Neurochem 55: 16

Dexter DT, Carayon A, Javoy-Agid F, Agid Y, Wells FR, Daniel SE, Lees AJ, Jenner P, Marsden CD (1991) Brain 114: 1953–1975

Dicker E, Cederbaum AI (1990) Generatios of reactive oxygen spedies and reduction of ferric chelates by microsomes in the presence of a reconstituted system containing ethanol, NAD+ and alcohol dehydrogenase. Alcohol Clin Exp Res 14: 238–244

Diezel PB (1955) Iron in the brain: a chemical and histochemical examination. In: Waelsch H (ed) Biochemistry of the developing nervous system. Academic Press, New York

Drayer B, Burger P, Hurwita B, Dawson D, Cain J (1987) Reduced signal intensity on MR images of thalamus and putamen in Multiple Sclerosis: increased iron content? AJNR 8: 413–419

Dwork AJ, Schon EA, Herbert J (1988) Nonidentical distribution of transferrin and ferric iron in human brain. Neuroscience 27: 333–345

Ehmann WD, Markesbery WR, Alauddin M, Hossain T, Brubaker EH (1986) Brain trace elements in Alzheimer's disease. Neurotoxicology 7: 197

Elovaara I, Icen A, Palo J, Erkinjuntti T (1985) CSF in Alzheimer' disease. J Neurol Sci 70: 73–80

Esiri MM, Taylor CR, Mason DY (1976) Applications of an immunoperoxidase method to a study of the central nervous system: preliminary findings in a study of human formalin-fixed material. Neuropathol Appl Neurobiol 2: 233–246

Espinosa A, Focaud B (1987) Effect of iron and transferrin on pure oligodendrocytes in culture; characterization of a high affinity transferrin receptor at different ages. Dev Brain Res 35: 123–130

Espinosa de los Monteros A, Kumar S, Scully S, Cole R, deVellis J (1990) Transferrin gene expression and secretion by rat brain cells in vitro. J Neurosci Res 18: 299–304

Fishman JB, Rubin JB, Handrahan JV, Connor JR, Fine RE (1987) Receptor mediated upkake of transferrin across the blood brain barrier. J Neurosci Res 18: 299–304

Fleming J, Joshi JG (1987) Ferritin: isolation of aluminum-ferritin complex from brain. Proc Natl Acad Sci USA 84: 7866

Francois C, Nguyen-legors J, Pencheron G (1981) Topographical and cytological localization of iron inrat and monkey brains. Brain Res 215: 317–322

Friden PM, Walus LR, Watson P, Doctrow SR, Kozarich JW, Backman C, Bergman H, Hoffer B, Bloom F, Granholm A–C (1993) Blood-brain-barrier penetration and in vivo activity of an NGF conjugate. Science 259: 373–377

Gelman BB, Rodrigua-Wolf, MS Wen J, et al (1992) Siderotic cerebral macrophages in the acquired immunodeficieny syndrome. Arch Pathol Lab Med 116: 509

Giometto B, Bozza F, Argentiero V, Gallo P, Pagni S, Piccinno MG, Tavolato B (1990) Transferrin receptor in rat central nervous system. An immunohistochemical study. J Neurol Sci 98: 81

Gocht A, Lohler J (1990) Changes in glial cell markers in recent and old demyelinated lesions in central pontine myelinolysis. Acta Neuropathol 80: 46–58

Good PF, Olanow CW, Perl DP (1992) Neuromelanin-containing neurons of the substantia nigra accumulate iron and aluminum in Parkinson's disease: a LAMMA study. Brain Res 593: 343–346

Goodman L (1953) Alzheimer's disease: a clinico-pathologic analysis of twenty-three cases with a theory on pathogenesis. J Nerv Ment Dis 118: 97

Graeber MB, Raivich G, Kreutzberg GW (1989) Increase in transferrin receptors and iron uptake in regenerating motor neurons. J Neurosci Res 23: 342–345

Hallgren B, Sourander P (1958) The effect of age on the non-haemin iron in the human brain. J Neurochem 3: 41–51

Hill JM, Ruff MR, Weber RJ, Pert CB (1985) Transferrin receptors in rat brain: neuropeptide-like pattern and relationship to iron distribution. Proc Natl Acad Sci USA 82: 4553–4557

Huggenvik JI, Craven CM, Idzerda RL, Bernstein S, Kaplan J, McKhight GS (1989) A splicing defect in the mouse transferrin gene leads to congenital atransferrinemia. Blood 74: 482

Jefferies WA, Brandon MR, Hunt SV, Williams AF, Gatter KC, Mason DY (1984) Transferrin receptor on endothelium of brain capillaries. Nature 312: 162

Kalaria RN, Sromek SM, Grahovac I, Harik SI (1992) Transferrin receptors of rat and human brain and cerebral microvessels and their status in Alzheimer's disease. Brain Res 585: 87–93

Kaneko Y, Kitamoto T, Tateishi J, Yamaguchi K (1989) Ferritin immunohistochemistry as a marker for microglia. Acta Neuropathol (Berl) 79: 129

Koeppen AH, Dentinger MP (1988) Brain hemosiderin and superficial siderosis of the central nervous system. J Neuropathol Exp Neurol 47: 249

Larkin EC, Rao A (1990) Importance of fetal and neonatal iron: adequacy for normal development of central nervous system. In: Dobking J (ed) Brain, behaviour, and iron in the infant diet, chapter 3. Springer, New York

Levine M, Tuil D, Uzan G, et al (1984) Expression of the transferrin gene during development of non-hepatic tissue. Biochem Biophys Res Comm 122: 212–217

Levine SM, Macklin WB (1990) Iron-enriched oligodendrocytes: a reexamination of their spatial distribution. J Neurosci Res 26: 508

Lin HH, Connor JR (1989) The development of the transferrin-transferrin receptor system in relation to astrocytes, MBP, and galactosecerebroside in normal and myelin-deficient rat optic nerves. Dev Brain Res 49: 281–293

Mash DC, Pablo J, Flynn DD, Efange SMN, Weiner WJ (1990) Characterization and distribution of transferrin receptors in the rat brain. J Neurochem 55: 1972

Morris CM, Candy JM, Oakley AE, Taylor GA, Mountfort S, Bishop H, Ward MK, Bloxham CA, Edwardson JA (1989) Comparison of the regional distribution of transferrin receptors and aluminum in the forebrain of chronic renal dialysis patients. J Neurol Sci 94: 295

Morris CM, Candy JM, Bloxham CA, Edwardson JA (1992) Distribution of transferrin receptors in relation to cytochrome oxidase activity in the human spinal cord, lower brainstem and cerebellum. J Neurol Sci 111: 158–172

Morris CM, Candy JM, Oakley AE, Bloxham CA, Edwardson JA (1992) Histochemical distribution of non-haem iron in the human brain. Acta Anat 144: 235–257

Nguyen-Legros J, Bizot J, Bolesse M, Publicani JP (1980) Noir de diamurobenzidine: une nouvelle methode histochimique de revelation du fer exogene. Histochemistry 66: 239

Nordmann R, Ribiere C, Rouach H (1990) Ethanol induced lipid peroxidation and oxidative stress in extrahepatic tissues. Alcohol 25: 231–237

Octave JN, Schneider YJ, Trouet A, Crichton RR (1983) Iron uptake and utilization by mammalian cells. 1. Cellular uptake of transferrin and iron. Trends Biochem Sci 8: 217

Oh TH, Markelonis GJ, Royal GM, Bregman BS (1986) Immunocytochemical distribution of transferrin and its receptor in the developing chicken nervous system. Dev Brain Res 30: 207

Partridge WM, Eisenberg J, Yang J (1987) Human blood brain barrier transferrin receptor. Metabolism 36: 892

Perry R (1986) Recent advances in neuropathology. Br Med Bull 42: 34

Rajan KS, Colburn RW, Davis JM (1976) Distribution of metal ions in the subcellular fractions of several rat brain areas. Life Sci 18: 423

Reinke LA, Rau JM, McCay PB (1990) Possible roles of free radicals in alcoholic tissue damage. Free Rad Res Comm 9: 205–211

Roskams AJ, Connor JR (1990) Aluminum access to the brain: a possible role for the transferrin receptor. Proc Natl Acad Sci USA 87: 9024

Roskams AJ, Connor JR (1992) The transferrin receptor in the myelin deficient (md) rat. J Neurosci Res 31: 421–427

Sofic E, Riederer P, Heinsen H, Beckmann H, Reynolds GP, Hebenstreit G, Youdim MBH (1988) Increased iron (III) and total iron content in post mortem substantia nigra of parkinsonian brain. J Neural Transm 74: 199

Taylor EM, Morgan EH (1990) Developmental changes in transferrin and iron uptake by the brain in the rat. Dev Brain Res 55: 35

Thompson CM, Marksberry WR, Ehmann WD, Mao Y-X, Vance DE (1988) Regional brain trace-element studies in Alzheimer's disease. Neurotoxicology 9: 1

Tsutsumi M, Sanders-Bush E (1990) 5-HT induced transferrin production by choroid plexus epithelial cells in culture: role of 5-HT receptor. J Pharmacol Exp Ther 254: 253–257

Yehuda S, Youdim MBH (1988) Brain iron deficiency: biochemistry and behavior. In: Yehude S, Youdim MBH (eds) Brain iron: neurochemical and behavioural aspects. Taylor and Francis, London

Correspondence: Dr. J. R. Connor, Department of Neuroscience and Anatomy, Pennsylvania State University, M. S. Hershey Medical Center, Hershey, PA 17033, U.S.A.

Iron deposits in brain disorders

K. Jellinger and E. Kienzl

Ludwig Boltzmann-Institute of Clinical Neurobiology, Lainz Hospital,
Vienna, Austria

Summary

Abnormal iron deposition in the CNS may be related to exogenous or endogenous factors. CNS is involved in idiopathic hemochromatosis, an autosomal recessively inherited storage disorder with increased resorption of iron in the gut. In superficial siderosis of the CNS resulting from repeated cerebral or subarachnoid bleeding, deposition of hemosiderin depends on accumulation of ferritin in the microglia. In experimental models, storage of heme-iron is mediated by shifts in the H/L ratio of ferritin. Tissue destruction is caused by Fe-catalyzed lipid peroxidation. Incrustation of neurons in perinatal brain injury or mitochondrial encephalopathies shows the presence of iron and calcium, also present in vascular deposits in basal ganglia and cerebellum in Fahr's disease, often related to hypoparathyroidism, hemochromatosis, and mitochondrial disorders. Iron deposition in the brain is found in various neurodegenerative disorders. In Hallervorden-Spatz disease, the globus pallidus and reticulata nigrae show iron-pigment deposits associated with axonals spheroids, increased iron content and uptake in the basal ganglia demonstrated in vivo by high-field MRI and PET. Decreased cystein dioxygenase in pallidum may result in accumulation of cystein that may locally bind iron and, via lipid peroxidation, may cause neuronal damage. In striatonigral degeneration, the putamen contains lipofuscin-like pigment with 4-fold increase in iron content suggesting pigment formation via lipid peroxidation. In Parkinson's disease, the damaged substantia nigra zona compacta shows increase of total iron and Fe^{3+}, mainly in neuromelanin, suggesting that iron-melanin interaction may be a pathogenic factor, while ferritin-reactive microglia indicates active degeneration. The pathogenic role of iron increased in neurofibrillary tangles in Alzheimer's disease needs further elucidation.

Introduction

Iron is an essential trace metal which plays a role in electron transport, enzymatic catalysis, cellular/neuronal development and cell damage

(Beard et al., 1993). In mammals, while there is no non-hem iron within the brain at birth, in adulthood it is concentrated only in a few regions, particularly the globus pallidus, substantia nigra, caudate/putamen, and cerebellar dentate nucleus (Hallgren and Sourander, 1958; Hill and Sitzer, 1984; Morris et al., 1992). These autopsy data on the distribution of non-hem iron in human brain were confirmed by high-field magnetic resonance tomography (MRI) (Drayer et al., 1986; Hall et al., 1992). Autochthonous brain iron is independent of hemoglobin and systemic iron metabolism. Due to the tightness of the blood-brain barrier for iron in adult brain, it is not influenced by experimental iron deficiency or systemic iron overload conditions (Dietzel and Taubert, 1954).

Abnormal Fe deposition in human brain has been reported in a number of acquired and genetic neurologic disorders and is considerad as a contributory factor to the pathogenesis of some neurode-degenerative diseases. Here, a brief review of pathologic Fe deposition in human brain of exogenous and endogenous origin is given.

Iron deposits of exogenous origin

Iron deposition related to exogenous causes is seen in idiopathic hemochromatosis, superficial hemosiderosis, in general paralysis, and as incrustation of neurons in damaged brain.

Idiopathic hemochromatosis

Iron deposition in the brain is seen in hemochromatosis, an autosomal recessively inherited Fe storage disorder encoded by a gene associated with particular HLA subtypes. Clinical features of the disease occurring after the age of 30 years, are liver cirrhosis, skin pigmentation, diabetes, hypogonadism, arthropathy, neuropathy, gait ataxia and dementia (Jones and Hedley-Whyte, 1983). Iron deposition due to increased resorption in the gut occurs in liver and in many organs causing liver cirrhosis, fibrosis of the pancreas and endocrine glands. Fe intoxication induces peroxidation of membrane lipids and damage to mitochondria causing cell necrosis. The CNS, in addition to multiple hemorrhages, shows brownish discoloration of the brain surface, choroid plexuses, hypothalamus, and area postrema, i.e. regions devoid of an intact blood-brain barrier (Cervos-Navarro, 1991). In addition to spongy changes and Alzheimer type II astroglia related to liver disease, there is an increase of lipofuscin (Miyasaki et al., 1977) and, in some cases, of physiological iron in the globus pallidus, reticulata nigrae, and dentate nucleus (Erbsloh, 1958). Iron is deposited in the cytoplasm of neurons and glial cells in lysosomes, associated with lipid bodies, or in perivascular macrophages (Cervos-Navarro, 1991).

Superficial siderosis of the CNS

A common effect of recurrent or persistent extravasation of blood into the subarachnoid space or the ventricles due to cerebral or subarachnoid hemorrhages is diffuse rust-brown staining of the exposed inner and outer surfaces of the brain and spinal cord. This superficial siderosis, an uncommon condition, is to be distinguished from CNS involvement in hemochromatosis. It may occur following repeated subarachnoid hemorrhages, around intracerebral hematomas or as a consequence of hemispherectomy, and may be associated with chronic neurological symptoms, progressive deafness and dementia (Koeppen and Dentinger, 1988; Stevens et al., 1991). Intravital recognition is possible by reduced T2-weighed signals in MRI (Bracchi et al., 1993). Histology shows hemosiderin deposits stained blue with Turnbull reaction in the meninges, adjacent CNS tissue and around cerebral blood vessels, accompanied by axonal swellings and glial proliferation (Fig. 1). Similar changes are

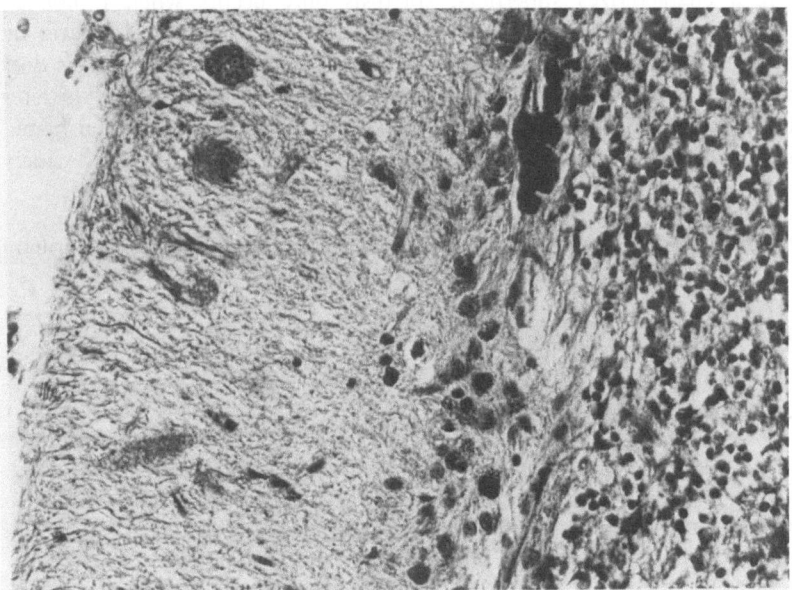

Fig. 1. Superficial siderosis of the CNS. Cerebellar cortex showing loss of Purkinje cells and swelling of their dendrites. H. & E. × 550

seen around old cerebral hemorrhages or vascular malformations (angiomas). Deposition of hemosiderin, i.e. iron derived from blood, depends on accumulation of ferritin, the most important iron storage protein, in the cytoplasm of microglia. Electron microscopy shows accumulation of ferritin granules only in the cytoplasm of microglia (Koeppen and Borke, 1991). The breakdown of hemoglobin and the release of iron from heme produces a local iron excess that stimulates ferritin synthesis by derepressing ferritin mRNA and dissociation of ferritin repressor protein (FRP) which are the principal mechanisms by which ferritin biosynthesis is controlled (Koeppen et al., 1992) In experimental animals, repeated intracisternal injection of autologous red blood cells causes accumulation of iron in the cytoplasm of microglial cells and astrocytes in the cerebral and cerebellar cortices. Immunocytochemistry for ferritin shows enhanced reaction product mainly in microglia, but hemosiderin occurs only after extending the injections of erythrocytes to 6 months (Koeppen and Borke, 1991). Biochemical studies by Koeppen et al. (1992) initially showed no increase of total iron or ferritin in the exposed cerebral areas. However, injections of red blood cells cause dramatic increase of the ratio of heavy (H) to light (L) isoforms of ferritin of 4:1, while in later stages a drop of the H/L ratio coincides with the appearance of granular hemosiderin (Table 1). These data suggest that incrustation by hemosiderin which characterizes superficial siderosis only occurs when prolonged exposure to hemoglobin produces persistent shifts of the H/L ratio by accumulation of L-ferritin which promotes cerebral iron storages, as does in liver. Iron storage is first in the form of Fe-ferritin, and is later followed by hemosiderin formation. The latter occurs only when heme-iron induces the formation of larger amounts of L-ferritin. It is widely

Table 1. Total iron and ferritin in normal and siderotic cerebellum and piriform cortex of the rabbit

Tissue	Duration of experiments	Fe (μg/g)	Ferritin (μg/g)	H/L ferritin
Cerebellum	control	20.6 ± 5.2	4.0 ± 0.1	1.36
	6 weeks	13.5 ± 3.3	4.3 ± 1.3	4.24
	3 months	15.5 ± 1.6	6.4 ± 3.2	2.06
	6 months	16.3 ± 3.8	3.8 + 1.4	0.42
Piriform cortex	control	22.5 ± 9.3	5.9 ± 0.2	1.25
	3 weeks	15.3 ± 2.9	7.3 ± 3.5	2.57
	3 months	17.6 ± 2.2	5.7 ± 1.5	1.84
	6 months	17.9 ± 7.8	4.1 ± 0.9	0.47

[Koeppen, et al (1992) J Neurol Sci 112: 38]

held that hemosiderin, showing a low iron content (8.5%), contains "degraded" ferritin, but it is not known to what degree H- and L-ferritin contribute to the "degraded" total ferritin in hemosiderin. Siderosis is usually well tolerated by the affected nervous tissue except for the cerebellar cortex showing damage to the molecular and Purkinje cells (Fig. 1). In the pathogenesis of the tissue destruction which accompanies CNS siderosis, hemosiderin may be less important than iron-catalyzed lipid perxidation caused by iron toxicity (Koeppen et al., 1992), and it appears of interest that only H-ferritin inhibits lipid peroxidation in vitro (Cozzi et al., 1990). There is no evidence that iron in brain hemosiderin is returned to the systemic circulation or is utilized for any further metabolic process in the brain.

Iron in general paralysis

In general paralysis (neurosyphilis) there is striking microglial prolife-ration in the cerebral cortex. The rod cells are easily demonstrated by Perls' stain ("paralysis iron") derived from killed treponemes and associa-ted with perivascular cuffing by lymphocytes and plasmocytes. These rod cells now gain retrospective importance in view of their reactivity for ferritin (Yoshioka et al., 1992).

Incrustation of neurons

Incrustation of nerve cells with hematophilic granules is found in old infarctions, in areas damaged during perinatal brain injury or in Leigh's subacute necrotizing encephalopathy, a mitochondrial encephalomyopa-thy due to various metabolic defects of the respiratory chain (Miranda et al., 1989). Necrotic neurons tend to remain in place after injury and present a mummified appearance. This type of lesion, known as "ferru-ginization" or "fossilization" of neurons, shows positive staining reactons for iron and calcium, the presence of which was confirmed by electron-probe analysis of the deposits (Leestma and Martin, 1968). The granules are located in neuronal cytoplasm, dendrites and axons, and in adjacent glial cells.

Calcification of basal ganglia (Fahr's disease)

A mild degree of mineralization in or around blood vessels in the stiratum is a common incidental finding in elderly brains. A more severe degree of calcification in the basal ganglia, visible on X ray (Fig. 2 a) and CCT is a frequent finding occurring in a variety of conditions (Huk et al., 1990). Calcification of far greater intensity and wider distribution, referred to as Fahr's disease, may occur in deficiency of parathyroid hormone, as familial disorder, or without demonstrable metabolic disorder. Patients may be asymptomatic or have seizures, abnormal movements, parkinsonism, ataxia, or dementia (Guseo, 1983; Cervos-Navarro, 1991). In this severe form, the mineralization is found in globus pallidus, striatum, cerebellar dentate nucleus, and other sites. It is localized to small and medium-sized vessels forming droplets or ring-like deposits that may form larger concretions, "brain stones". The surrounding nervous tissue may be intact or is destroyed, with very little glial reaction. The concretions are composed of amorphous material or crystals and form small lumps at the inner surface of the vessels (Fig. 2 b). Chemical and EDX-analysis have shown proteins, polysaccharides and metallic ions, in particular calcium, iron, magnesium, phosphorus, zinc, and aluminum (Duckett et al., 1977; Cervos-Navarro, 1991). The pathogenesis of mineralization is unknown; in 60% it results from hypoparathyroidism or other disturbances of calcium or phosphate metabolism; it may also be associated with disorders of mitochondrial metabolism – mitochondrial encephalomyopathies (Truong et al., 1990) and occurs in families with hemochromatosis, while in other cases the causes of CNS mineralization are unknown.

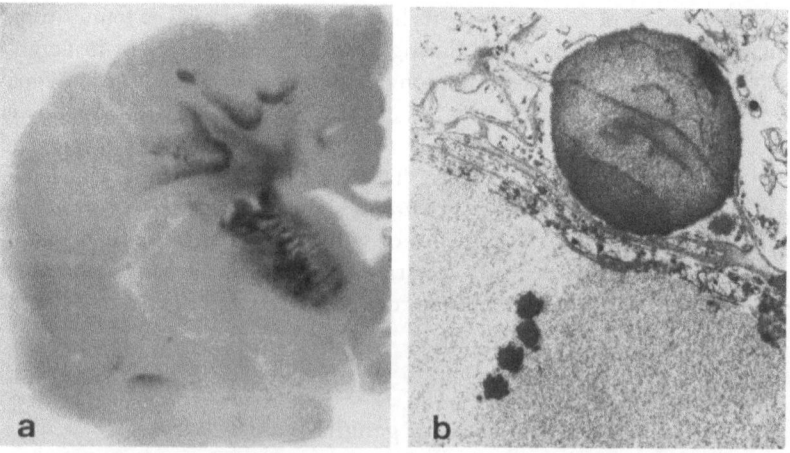

Fig. 2. a Fahr's disease. Post-mortem X-ray shows multiple calcifications in the basal ganglia. b Amorphous concretions in the inner layers of cerebral vessels

Iron deposits in neurodegenerative disorders

Deposition of iron and iron-containing pigments in basal ganglia are found in a number of neurodegenerative disorders, e. g. Hallervorden-Spatz disease, striatonigral degeneration, Parkinson's and Huntington's disease, while selective accumulation of iron & aluminum has been reported in neurofibrillary tangles in Alzheimer's disease (Good et al., 1992a).

Hallervorden-Spatz disease (HSD)

Deposits of iron-containing pigment in the pallidum and reticulata nigrae are found in HSD, a rare, often familial (partly autosomal dominant) degenerative extrapyramidal motor disorder manifested by choreoathetosis, rigidity, dystonia, retinal degeneration, and progressive dementia (Jellinger, 1992; Tripathi et al., 1992). Pathologically, there is a rust-brown discoloration of the globus pallidus and reticulata nigrae showing loss of neurons, gliosis, iron-pigment deposits and axonal spheroids (Fig. 3) also occurring elsewhere in the CNS. The yellow-brownish pigment may be intracellular in neurons, astro- and microglia or may lie free in the neuropil, usually around blood vessels. It shows strong Perls' reaction for Fe^{3+}, positive PAS reaction and sudanophilia, indicating a mixture of iron (acid hematin) lipofuscin and neuromelanin. Biochemical studies show a 3–4 fold increase of iron in the putamen and pallidum (Vakili et al., 1977; Swaiman, 1991) and an increase uptake of radio-

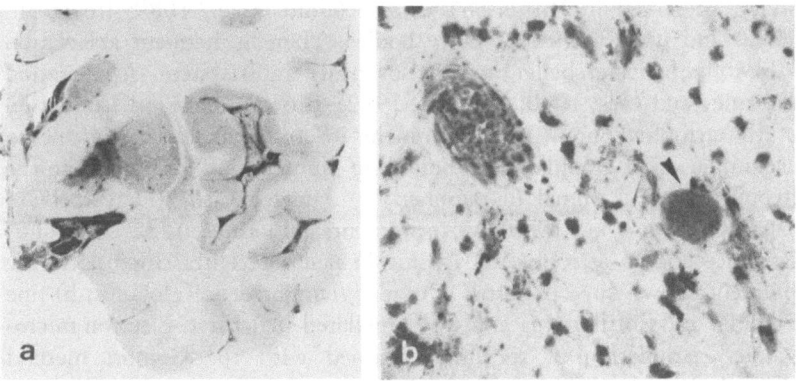

Fig. 3. Hallervorden-Spatz disease. a Discoloration of globus pallidus. b Iron pigment and axonal spheroids (arrow) in zona reticulata nigrae; cresyl violet × 800

active Fe in the basal ganglia (Vakili et al., 1977; Leenders, 1993), without signs of systemic disorders of iron metabolism (Swaiman, 1991). Decreased T2-weighed intensity in pallidum and substantia nigra indicating increased iron and ferritin ("tiger-eye phenomenon") using high-field MRI allow the in vivo diagnosis of HSD (Sethi et al., 1988; Angelini et al., 1992; Savoiardo et al., 1993).

Recent reports of decreased activity of the enzyme cystein dioxygenase in the globus pallidus of patients with HSD (Perry et al., 1985) may result in accumulation of cystein that, together with ferrous ions, stimulates lipid peroxidation in microsomes (Searle and Wilson, 1983). High levels of iron and increased cystein concentration in the pallidum generate reactive peroxides which, via a chain reaction, can damage the highly saturated fatty acids in neuronal membranes. Perhaps the accumulation of cystein acts as a local chelating agent which binds iron (Jameson and Linert, 1991). This may explain the increased iron uptake by the pallidum in HSD (Vakili et al., 1977; Swaiman, 1991) and both the formation of pigment and spheroids in this disorder by peroxidation (Park et al., 1975).

Striatonigral degeneration (SND)

SND is a rare neurodegenerative disorder clincially characterized by atypical parkinsonism, dystonia, orthostatic hypotension, cerebellar symptoms, and dementia with failure to respond to L-dopa (Fearnley and Lees, 1990; Jellinger, 1992). Morphology shows marked atrophy of the putamen without sparing of the large striatal neurons and of substantia nigra zona compacta with loss of calcineurion and calbindin immunoreactivity, preserved in Parkinson's disease (Goto et al., 1989; Ito et al., 1992) and usually without Lewy bodies. There is frequent association with olivopontocerebellar atrophy or other multisystem degeneration (Fearnley and Lees, 1990; Jellinger, 1992). Brown-greenish discoloration of the atrophic putamen, a prominent macrosocpic feature, is due to accumulation of granular iron containing pigment and is reflected in the hypointense T 2 signal in putamen in high-field MRI (Drayer et al., 1986; O'Brien et al., 1990). Most pigments are located in the extracellular space and in astroglia, only some in neurons. Ultrastructurally, the pigment shows three patterns: a) coarse granular dense globules, b) fine granular and fibrillary materials. c) lamellated structures. Electron microscopic examination of specimens stained with the Gomori method showed that the iron reaction was present in the granular pigments (Fig. 4). Determination of trace metals by atomic absorption spectroscopy revealed that the mean putamen iron content in SND was about four

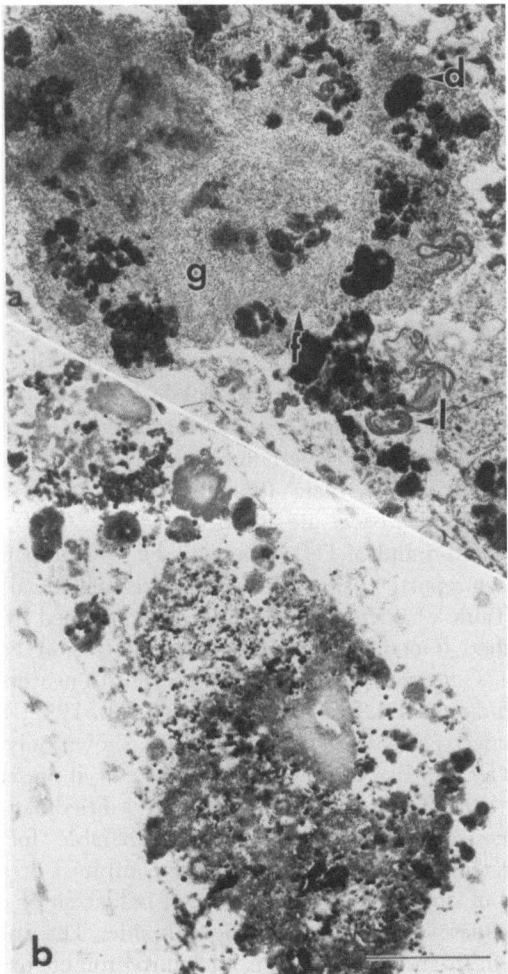

Fig. 4. Striatonigral degeneration. **a** Electron micrograph of granular pigment in the putamen showing coarse, electron dense globules (*d*), fine granular (*g*) and fibrillary materials (*f*), and lamellated structures (*l*). **b** Electron micrograph of pigment stained for iron with Gomori method. Electron dense reaction products denoting presence of iron can be seen

times greater than in controls, while copper content was significantly lower (Kato et al., 1992). The demonstration of the striking increase of iron in the putamen in SND and its deposition in the abnormal pigment that ultrastructurally resembles that in HSD suggest that the abnormally high iron concentration could be related to pigment formation, possibly due to lipid peroxidation reactions initiated by Fe (Kato et al., 1992). Although the origin of the pigment in both HSD and SND is unknown, there could be some common pathogenic factors for pigment formation and neurodegeneration in both disorders.

Parkinson's disease (PD)

The mechanism that leads to accelerated degeneration of melanized nigrostriatal dopaminergic neurons in the substantia nigra zona compacta (SNZC) are still unknown. There is evidence that subpopulations of dopaminergic neurons in the midbrain are selectively vulnerable, i.e. the highly melanized ventral tier of area A9 lacking calbindin immuno-reactivity (Hirsch, 1992) suggesting a neuroprotective role of the enzyme due to prohibition of cellular calcium influx. The SN is rich in Fe, particularly the zona reticulata (SNZR), where dendrites from the dopaminergic zona compacta are localized. In SNZC of PD brain, an increase of total iron and of Fe^{3+} – but not Fe^{2+} – compared to control subjects has been reported using biochemical, histochemical, and physical methods (see Table 2). Increase of Fe has been confirmed by the sensitive LAMMA method (Good et al., 1992b), an increase of Fe^{3+} in neuro-melanin of SNZC neurons of PD patients, but not in neuronal cytoplasm, Lewy bodies and neuropil, by EDX (Jellinger et al., 1992, 1993; Fig. 5), while Moessbauer spectroscopy of total SN has shown only the presence of ferric (Fe^{3+}) ferritin like iron, while ferrous (Fe^{2+}), if present at all, may represent less than 10% of the total iron (Galazka-Friedman et al., 1993).

What makes the SNZC particularly vulnerable for a potential damaging effect of any increase in Fe is the combined presence of high concentrations of dopamine and high rate of oxidation of dopamine via monoamine oxidase to form hydrogene peroxide, The combination of oxygen free radicals and iron, which is known to promote oxidative stress, causes damage to biological membranes, cellular DNA, mitochondrial function, etc, leading to cell destruction (Fahn and Cohen, 1992; Jenner et al., 1992; Sofic et al., 1992; Uney et al., 1992; Youdim et al., 1993). With iron overload, oxidative stress may contribute selectively to the death of melanized dopaminergic neurons in the SNZC of PD patients (Fig. 6). However, the mechanisms which account for iron uptake in neuromelanin of dopaminergic SN cells remains unknown, since regional binding intensity of transferrin receptors, being important for Fe cell uptake from transferrin (Aisen, 1992), according to recent studies by Faucheux et al. (1993), is not different between PD and control brains, suggesting that iron in SN does not accumulate via increase of these receptors in this area. While Dexter et al. (1990), using a RIA method

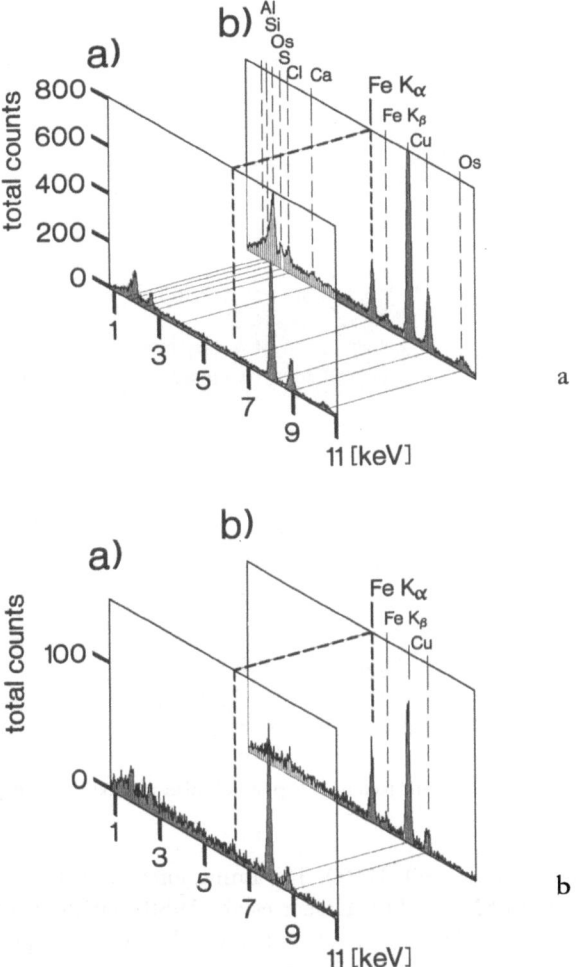

Fig. 5. a Determination of trace metals in substantia nigra with EDX analysis: *a* Iron (Fe) in Lewy body not detectable [f (FeKα) = 0.79], *b* in neuromelanin increased [f (FeKα) = 0.0]. **b** Comparison of Fe contents in *a* synthetic dopamine-melanin [f (FeKα = 0.5], *b* in synthetic Fe^{3+} melanin complex showing same spectra as neuromelanin in patients with Parkinson's disease [f (FeKα) = 0.0]

with a monoclonal antibody (mAB) against ferric L-ferritin, found decreased levels in SN and other areas of PD brain, immunocytochemical studies with a mAB against human liver ferritin consisting of both L + H subunits, showed increase of ferritin immunoreactive microglia in SN of both PD and PD-AD brain, while these cells were absent in controls

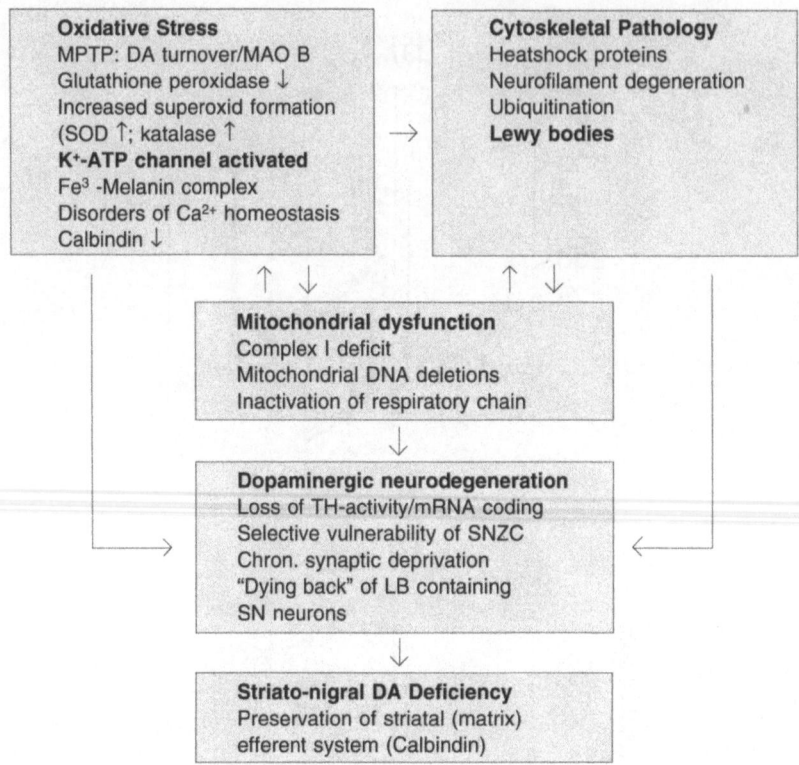

Fig. 6. Major pathogenic factors in Parkinson's disease (modified from Jellinger, 1993)

(Jellinger et al., 1990, 1993). The immunoreactions for both ferritin and HLA-DR in SN of PD (McGeer et al., 1988) outline reactive microglia indicating an active degenerating process. Microglia, responding to cellular degeneration in the SNZC that occurs also after nigral iron infusions in rats (Sengstock et al., 1993) may themselves secrete free radicals and thus exacerbate oxidative damage to SN neurons. The selective increase of Fe (III) and of ferritin-positive microglia in SNZC of PD brain is thought not to be the consequence of but a contributory factor of neuronal death (Ben-Shachar and Youdim, 1993), the definite causes of which remain to be elucidated.

Iron in Alzheimer's disease

In AD brain, in addition to increased aluminum concentrations (Dedman et al., 1992), recent studies using the highly sensitive LAMMA-laser

Table 2. Iron content in substantia nigra of PD patients found by different investigators and using different methods

Authors	Method	Fe tot.	Fe^{3+}	Fe^{2+}	Region
Sofic et al. (1988)	SP	3+	3+	±	SN
Sofic et al. (1991)	SP	3+	3+	±	SNC
		±	±	±	SNR
Dexter et al. (1989)	CPS	3+			SN
Dexter et al. (1991)	CPS	3+			SN
Uitti et al. (1989)	AA/AES	±			SN
Jellinger et al. (1990)	PIH		3+	−	SNC
			±	−	SNR
Hirsch et al. (1991)	EDX	3+			SN
		3+			LB
Jellinger et al. (1992, 1993)	EDX	3+	3+	−	SNC
		±	±	−	LB
Perl et al. (1992)	LAMMA	3+			SN
Galazka-Friedman (1993)	MS	3+	3+	±	SN
Rutlege (1988)	HFMRI	2+			SN
		2+			PUT
DeVolder et al. (1989)	HFMRI	3+			PUT

3+ Increased compared to controls; ± unchanged; − not detectable; *SN* substantia nigra; *SNC, SNR* pars compacta, reticulata; *LB* Lewy body; *PUT* putamen; *SP* spectrophotometry; *CPS* coupled plasma spectroscopy; *PIH* Perl's histochemistry; *EDX* energy dispersive X-ray microanalysis; *MS* Mössbauer spectroscopy; *LAMMA* Laser microprobe mass analysis; *AA/ AES* atomic adsorption/atomic emission spetroscopy; *HFMRI* high field magnetic resonance tomography

microprobe mass analysis, have demonstrated significantly elevated iron levels in the neurofibrillary tangles (Good et al., 1992a). In addition, iron accumulates around and within senile plaques (Connor et al., 1992), while they contain no aluminum (Chafi et al., 1991). The nature and consequence of the selective accumulation of iron and aluminum, both highly reactive and potentially toxic metals, is unknown. Fibrillary tangles are composed of an abnormally phosphorylated form of the microtubule-associated tau protein that later undergoes ubiquitination (Bancher et al., 1991). Binding of metal ions to the tau molecule may cause blockage of tau binding inducing an abnormal accumulation of tau within the neuronal perikarya. The presence of iron along with aluminum also suggests that transferrin, the major carrier for iron, may play a role in tangle formation. Transferrin has unoccupied metal binding sites that may allow aluminum to be carried by transferrin, presumably with incorporation into cellular compartments mediated by transferrin receptors (Roskams and Connor, 1986; Aisen, 1992). In addition, transferrin is increased around senile plaques, most of the ferritin-containing cells associated with plaques and blood vessels being microglia (Jellinger et al., 1990; Connor et al., 1992) that play an important role in the processing or deposition of β-amyloid and plaque formation in AD (Wisniewski et al., 1992). The extracellular localisation of diffuse iron and transferrin in and around plaques suggests extravasation or active phagocytosis of extravasated iron and/or active synthesis of ferritin. These data suggest a disruption of brain iron homeostasis in AD as demonstrated by alterations in the normal cellular distribution of iron and the proteins responsible for iron regulation (Connor et al., 1992). These changes causing increased iron concentrations in AD brain may contribute to Fe-induced free radical or lipid peroxidation, suggested by some authors as a pathogenic factor of neuronal degeneration in AD (Richardson et al., 1992), which is in line with the radical theory of aging (Harman, 1992).

In conclusion, abnormal iron deposition in the CNS related to both exogenous and endogenous or unknown factors, through a variety of biochemical and biophysical processes, may contribute to neuronal damage and progressive degeneration leading to a variety of neuro logical disorders, the pathogensis of most of which remains to be elucidated.

References

Aisen P (1992) Entry of iron into cells: a new role for the transferrin receptor modulating iron release from transferrin. Ann Neurol 32 [Suppl]: S62–S68

Angelini L, Nardocci N, Rumi V, Zorzi C, Strada L, Savoiardo (1992) Hallervorden-Spatz disease. Clinical and MRI study of 11 cases diagnosed in life. J Neurol 239: 417–425

Bancher C, Grundke-Iqbal K, Iqbal K, Fried VA, Smith HT, Wisniewski HM (1991) Abnormal phosphorylation of tau precedes ubiquitination in neurofibrillary pathology of Alzheimer disease. Brain Res 539: 11–18

Beard JL, Connor JC, Jones BC (1993) Iron in the brain. Nutr Rev 51: 157–170

Ben-Shachar D, Youdim MBH (1993) Iron, melanin and dopamine interaction: relevance to Parkinson's disease. Prog Neuropsychopharmacol Biol Psychiatry 17: 139–150

Bracchi M, Savoiardo M, Triulzi F, Daniele D, Grisoli M, Bradac GB, Agostinis C, Pelucchetti D, Scotti G (1993) Superficial siderosis of the CNS: MR diagnosis and clinical findings. AJNR 14: 227–236

Cervos-Navarro J (1991) Pathologie des Nervensystems V. Degenerative und metabolische Erkrankungen. In: Doerr W, Uehlinger E (Hrsg) Spezielle pathologische Anatomie. Springer, Berlin Heidelberg New York Tokyo, S 419–462

Chafi AH, Hauw JJ, Rancurel G, Berry JP, Galle C (1991) Absence of aluminium in Alzheimer's disease brain tissue: electron microprobe and ion microprobe studies. Neurosci Lett 123: 61–64

Connor JR, Menzies SL, St. Martin SM St, Mufson EJ (1992) A histochemical study of iron, transferrin, and ferritin in Alzheimer's disease brains. J Neurosci Res 31: 75–83

Dedman DJ, Treffry A, Candy JM, Taylor GAA, Morris CM, Perry RH, Edwardson JA, Harrison PM (1992) Iron and aluminium in reation to brain ferritin in normal individuals and Alzheimer's disease and chronic dialysis patients. Biochem J 287: 509–514

De Volder AG, Francart J, Laterre C, et al (1989) Decreased glucose utilization in the striatum and frontal lobe in probable striatonigral degeneration. Ann Neurol 26: 239–247

Dexter TD, Wells FR, Agid F, Agid Y, Lees A, Jenner P, Marsden CD (1989) Increased nigral iron content and alteration in other metals occuring in Parkinson's disease. J Neurochem 52: 1830–1836

Dexter TD, Carayon A, Javoy-Agid F, et al (1991) Alterations in the levels of iron, ferritin and other trace metals in Parkinson's disease and other neuro-degenerative disease affecting the basal ganglia. Brain 114: 1953–1975

Diezel PB, Taubert M (1954) Untersuchungen am Gehirneisen. Verh Dtsch Ges Path 38: 221–225

Drayer BP, Olanow W, Burger P, Johnson GA, Herfkens S, Riederer S (1986) Parkinson plus syndrome: diagnosis using high field MR imaging of brain iron. Radiology 159: 493–498

Duckett S, Galle P, Escouroll R, Poirier J, Hauw JJ (1977) Presence of zinc, aluminium, magnesium in striatopallidodentate (SPD) calcifications (Fahr's disease): electron probe study. Acta Neuropathol 38: 7–10

Erbslöh F (1958) Kernikterus. In: Lubarsch 0, Henke F, Rossle R (Hrsg) Handbuch der speziellen pathologischen Anatomie und Histologie, Bd XIII/2. Springer, Berlin Göttingen Heidelberg, S 1769

Fahn S, Cohen G (1992) The oxidative stress hypothesis in Parkinson's disease: evidence supporting it. Ann Neurol 32: 804–812

Faucheux BA, Hirsch EC, Villares J, Selimie F, Mouatt-Prigent A, Javoy-Agid F, Hauw JJ, Agid Y (1993) Distribution of (^{125}I)-ferrotransferrin binding in the mesencephalon of control subjects and patients with Parkinson's disease. J Neurochem (in press)

Fearnley JM, Lees AJ (1990) Striatonigral degeneration. Brain 113: 1823–1842

Galazka-Friedman J, Bauminger ER, Friedman A, Barcikowska M, Sualski J, Hechel D (1993) The role of iron in Parkinson's disease: Mossbauer spectroscoy study. J Neurochem (in press)

Good PF, Perl DP, Bierer LM, Schmeidler J (1992a) Selective accumulation of aluminum and iron in the neurofibrillary tangles of Alzheimer's disease: a laser microprobe (LAMMA) study. Ann Neurol 31: 286–292

Good PF, Olanow WC, Perl DP (1992b) Neuromelanin-containing neurons of the substantia nigra accumulate iron and aluminium in Parkinson's disease. A LAMMA study. Brain Res 593: 343–346

Goto S, Hirano A, Matsumoto S (1989) Subdivisional involvement of nigrostriatal loop in idiopathic Parkinson's disease and striatonigral degeneration. Ann Neurol 26: 766–770

Guseo A (1983) Striopallidodentale Verkalkungen (Fahrsches Syndrom). In: Hopf

H Ch, Poeck K, Schliack H (Hrsg) Neurologie in Praxis und Klinik, Bd I.
  G Thieme, Stuttgart, S 6.25–6.29
Hall S, Rutledge JN, Schallert T (1992) MRI, brain iron and experimental
  Parkinson's disease. J Neurol Sci 111: 198–208
Hallgren B, Sourander P (1958) The effect of age on the nonhaem iron in the
  human brian. J Neurochem 3: 41–51
Harman D (1992) Free radical theory of aging: a hypothesis on pathogenesis of
  senile dementia of the Alzheimer type. Age 16: 23–30
Hill JM, Switzer RC (1984) The regional distribution and cellular localization of
  iron in the rat brain. Neuroscience 11: 595–603
Hirsch EC (1992) Why are nigral catecholaminergic neurons more vulnerable
  than other cells in Parkinson's disease? Ann Neurol 32: S88–S93
Hirsch EC, Brandel J-P, Galle P, Javoy-Agid F, Agid Y (1991) Iron and
  aluminum increase in the substantia nigra of patients with Parkinson's disease.
  An X-ray microanalysis. J Neurochem 56: 446–451
Huk WJ, Gademann G, Friedmann G (1990) MRI of central nervous system
  diseases. Functional anatomy, imaging, neurological symptoms, pathology.
  Springer, Berlin Heidelberg New York Tokyo, pp 215–216
Ito H, Goto S, Sakamoto S, Hirano A (1992) Calbindin-D28k in the basal ganglia
  of patients with parkinsonism. Ann Neurol 32: 543–550
Jameson RF, Linert W (1991) Complex formation folowed by internal electron
  transfer: the reaction between cysteine and iron (III). Monatsh Chemie 122:
  887–906
Jellinger K (1992) Sonderformen; Progressive Pallidumatrophie, Hallervorden-
  Spatz-Krankheit, striatonigrale Degeneration. In: Hopf H Ch, Poeck K,
  Schliack H (Hrsg) Neurologie in Praxis und Klinik, Bd II, 2. Aufl. Thieme,
  Stuttgart, S 4.20–4.27
Jellinger K (1993) Pathogenese und Pathophysiologie der Parkinson-Krankheit.
  Neuropsychiatrie 7: 29–37
Jellinger K, Paulus, W, Grundke-Iqbal I, Riederer P, Youdim MBH (1990)
  Brain iron and ferritin in Parkinson's and Alzheimer's diseases. J Neural
  Transm [P-D Sect] 2: 327–340
Jellinger K, Kienzl E, Rumplmair G, Riederer P, Stachelberger D, Ben-Shachar
  D, Youdim MBH (1992) Iron-melanin complex in substantia nigra of Parkin-
  sonian brains: an X-ray microanalysis. J Neurochem 59: 1168–1172
Jellinger K, Kienzl E, Rumplmair G, Paulus W, Riederer P, Stachelberger H,
  Youdim MBH, Ben Shachar D (1993) Iron and ferritin in substantia nigra in
  Parkinson's disease. Adv Neurol 60: 267–272
Jenner P, Schapira AHV, Marsden CD (1992) New insights into the cause of
  Parkinson's disease. Neurology 42: 2241–2250
Jones HR Jr, Hedley-Whyte ET (1983) Idiopathic hemochromatosis (IHC).
  Neurology 33: 1479–1483
Kato S, Meshitsuka S, Ohama E, Tanaka J, Llena JF, Hirano A (1992) Increased
  iron content in the putamen of patients with striatonigral degeneration. Acta
  Neuropathol 84: 328–330
Koeppen AH, Dentinger MP (1988) Brain hemosiderin and superficial siderosis
  of the central nervous system. J Neuropathol Exp Neurol 47: 249–270

Koeppen AH, Borke RC (1991) Experimental superfical siderosis of the central nervous system. I. Morphological studies. J Neuropathol Exp Neurol 50: 579–594

Koeppen AH, Hurwitz CG, Dearborn RE, Dickson AC, Borke RC, Chu RC (1992) Experimental siderosis of the central nervous system: biochemical correlates. J Neurol Sci 112: 38–45

Leenders KL (1993) Personal communication

Leestma JE, Martin E (1968) An electron probe and histochemical study of ferru ginated neurons. Arch Pathol 122: 597–605

Miranda AF, Ishii S, DiMauro S, Shay JW (1989) Cytochrome oxidase C deficiency in Leigh's syndrome. Neurology 39: 697–702

McGeer PL, Itagaki S, Boyes B, McGeer EG (1988) Reactive microglia are positive for HLA-DR in the substantia nigra of Parkinson's and Alzheimer's disease brains. Neurology 38: 1285–1291

Miyasaki K, Muaro S, Koizumi N (1977) Hemochromatosis associated with brain le sions: a disorder of trance metal binding proteins and/or polymers? J Neuropathol Exp Neurol 36: 964–976

Morris CM, Candy JM, Oakley AE, Bloxam CA, Edwardson JA (1992) Histochemical distribution of hon-haem iron in the human brain. Acta Anat 144: 235–257

O'Brien C, Sung JH, McGeachie RE, Lee MC (1990) Striatonigral degeneration; clinical, MRI, and pathologic correlation. Neurology 40: 710–711

Park BE, Netsky MG, Bertsill WL (1975) Pathogenesis of pigment and spheroid formation in Hallervorden-Spatz syndrome and related disorders. Neurology 25: 1172–1178

Perl DP, Good PF (1992) Comparative techniques for determinating cellular iron distribution in brain tissues. Ann Neurol 32 [Suppl]: S76–S81

Perry TL, Norman MG, Yong VW, Whiting S, Crichton JU, Hansen S, Kish SJ (1985) Hallervorden-Spatz disease: cystein accumulation and cystein dioxygenase deficiency in the globus pallidus. Ann Neurol 18: 482–489

Riederer P, Dirr A, Goetz M, Sofic E, Jellinger K, Youdim MBH (1992) Distribution of iron in different brain regions and subcellular compartments in Parkinson's disease. Ann Neurol 31 [Suppl]: S101–S104

Richardson JS, Subbarao KF, Ang LC (1992) On the possible role of iron-induced free radical peroxidation in neuronal degeneration in Alzheimer's disease. Ann NY Acad Sci 648: 326–327

Roskams AJ, Connor JR (1990) Aluminum access to the brain: a role for transferrin and its receptor. Proc Natl Acad Sci USA 87: 9024–9027

Savoiardo H, Halliday WC, Nardocci N, et al (1993) Hallervorden-Spatz disease: MR and pathologic findings. Am J Neuroradiol 14: 155–162

Searle AJF, Willson RL (1983) Stimulation of microsomal lipid peroxidation by iron and cysteine: characterization and the role of free radicals. Biochem J 212: 549–554

Sengstock GJ, Olanow CW, Menzies RA, Dunn AJ, Arendash GW (1993) Infusion of iron into the rat substantia nigra: nigral pathology and dose-dependent loss of striatal dopaminergic markers. J Neurosci Res 34 (in press)

Sethi KD, Adams RJ, Loring DW, El Mammal T (1988) Hallervorden-Spatz syndrome: magnetic resonance imaging correlations. Ann Neurol 24: 692–694

Sofic E, Riederer P, Heinsen H, Beckann H, Reynolds GP, Hebenstreit G, Youdim MBH (1988) Increased iron III and total iron content in postmortem substantia nigra in parkinsonian brains. J Neural Transm 74: 199–205

Sofic E, Paulus W, Jellinger K, Riederer P, Youdim MBH (1991) Selective increase of iron in substantia nigra zona compacta of parkinsonian brain. J Neurochem 56: 987–982

Sofic E, Lange KW, Jellinger K, Riederer P (1992) Reduced and oxidized glutathione in the substantia nigra of patients with Parkinson's disease. Neurosci Lett 142: 128–130

Stevens I, Petersen D, Grodd W, Poremba M, Dichgans J (1991) Superficial siderosis of the central nervous system. Arch Psych Clin Neurol 241: 57–60

Swaiman KF (1991) Hallervorden-Spatz syndrome and brain iron metabolism. Arch Neurol 48: 1285–1294

Tripathi RC, Tripathi BJ, Baserman SC, Park JK (1992) Clinicopathologic correlation and pathogenesis of ocular and central nervous system manifestations in Hallervorden-Spatz syndrome. Acta Neuropathol 83: 113–119

Uitti RJ, Rajput AH, RozdilskY B, Bickis M, Wollin T, Yuen WK (1989) Regional metal concentrations in Parkinson's disease and control brains. Can J Neurol Sci 16: 310–31

Uney JB, Anderton BH, Thomas SM (1992) Changes in heath shock protein 70 and ubiquitin mRNA levels in Cl300 N2A mouse neuroblastoma cells following treatment with iron. J Neurochem 60: 659–665

Vakili S, Drew AL, von Schuching S, Becker D, Zedman W (1977) Hallervorden-Spatz syndrome. Arch Neurol 34: 729–738

Wisniewski HM, Wegiel J, Wang KH, Lach B (1992) Ultrastructural studies of the cells forming amyloid in the cortical vessel wall in Alzheimer's disease. Acta Neuropathol 84: 117–127

Yoshika M, Shapshak P, Sun WCJ, Nelson SJ, Svenningsson A, Tate LG, Pardo V, Resnik L (1992) Simultaneous detection of ferritin and HIV-1 in reactive microglia. Acta Neuropathol 84: 297–306

Youdim MBH, Ben-Shahar D, Riederer P (1993) The possible role of iron in the etiopathology of Parkinson's diseases – review. Mov Disord 8: 1–12

Correspondence: Prof. Dr. K. Jellinger, Ludwig Boltzmann-Institute of Clinical Neurobiology, Lainz-Hospital, Wolkersbergenstrasse 1, A-1130 Wien, Austria.

Brain iron and schizophrenia

K. W. Lange[1], J. Kornhuber[1], P. Kruzik[2], W.-D. Rausch[2],
E. Gabriel[3], K. Jellinger[4], and P. Riederer[1]

[1] Department of Psychiatry, University of Würzburg, Würzburg,
Federal Republic of Germany
[2] Institute of Medical Chemistry, University of Veterinary Medicine,
[3] Psychiatric Hospital Baumgartnerhöhe, and [4] Ludwig-Boltzmann-Institute
of Clinical Neurobiology, Vienna, Austria

Summary

The concentration of iron was determined by atomic absorption spectroscopy
in post-mortem tissue from various brain regions in schizophrenic patients and
control subjects without neuropsychiatric diseases. Analysis of iron content
showed a clear regional difference with highest iron levels in the caudate nucleus.
There was no correlation between iron content and the neuroleptic-free period
prior to death. Iron content in the schizophrenic group was not different from
controls for the cortex, gyrus cinguli, caudate nucleus, hippocampus, amygdala,
corpus mamillare and hypothalamus. In one patient who had suffered from
tardive dyskinesia prior to death, the iron concentration in all the brain regions
examined was within the mean ± 2 SD range of those schizophrenic subjects
without prior tardive dyskinesia. The present results suggest that there are no
profound differences in the content of iron in post-mortem brain tissue of
schizophrenic and control subjects.

Introduction

Iron is the most abundant metal in the human body (Pollitt and
Leibel, 1982) and the brain contains a substantially higher concentration
of iron than of any other metal (Youdim, 1988). Within the brain, iron
shows an uneven distribution with high levels in the globus pallidus, red
nucleus, substantia nigra, putamen, dentate nucleus and caudate nucleus
(Spatz, 1924; Hallgren and Sourander, 1958; Hill and Switzer, 1984;

Riederer et al., 1989). Although the function of a regionally high brain iron content is unknown the homoeostasis of brain iron is thought to be necessary for normal brain function (Youdim et al., 1989; Youdim, 1990). The regional distribution of iron within the brain is important because areas with a high iron content play a role in the pathogenesis of movement disorders and also have high concentrations of dopamine, γ-aminobutyric acid, serotonin and neuropeptides (Hill, 1988) which have been related to movement disorders and psychiatric diseases.

Iron plays an important role as a catalyst in oxidation, hydroxylation and peroxidation reactions. Due to high iron concentrations the basal ganglia are particularly susceptible to iron-mediated lipid peroxidation and iron may play a part in neurotoxic processes involved in the pathophysiology of neurodegenerative diseases such as Parkinson's disease (Lange et al., 1992).

Transition metals including iron have been linked with schizophrenia (for review see Kornhuber et al., 1993). Basal ganglia mineralization has been shown to manifest itself clinically as a schizophreniform disorder in a substantial number of patients (Lowenthal and Bruyn, 1968). Some early studies reported a subjective impression of increased iron staining in post-mortem brain sections from schizophrenic patients (Josephy, 1930; Hopf, 1952; Stevens, 1982). Since the deposition of iron in nervous tissue occurs simultaneously with that of other minerals, including calcium, magnesium and zinc (Schiffer, 1971; Smeyers-Verbeke et al., 1975), the results of previous clinico-pathological correlation studies with regard to iron may have been confused by other elements. Studies using quantifying methods specific for iron detection are needed in order to assess the role of iron in the pathogenesis of schizophrenia. The present study has examined the content of iron, using atomic absorption spectroscopy, in various brain regions of patients who had died with schizophrenia in comparison with control subjects.

Methods and materials

Brain tissue was obtained at autopsy from 11 schizophrenic patients (6 female, 5 male, mean age ± S.D.: 69.6 ± 8.2 years, age range 57–80 years) diagnosed according to both Feighner et al. (1972) and the International Classification of Diseases (ICD-9). The diagnostic subgroups according to the ICD were schizophrenia simplex (ICD 295.0, n = 1), hebephrenic subtype (ICD 295.1, n = 1), paranoid subtype (ICD 295.3, n = 1), chronic schizophrenia (ICD 295.6, n = 7) or schizoaffective psychosis (ICD 295.7, n = 1). All schizophrenic patients had received neuroleptic medication prior to death, three patients had been drug-free for at least one year and seven patients for at least three months. According to the case notes one female patient aged 77 years had developed tardive dyskinesia.

Control tissue was obtained from 12 subjects (10 female, 2 male, average age 75.3 ± 7.1 years, age range 41–91 years) with no history of neurological or psychiatric diseases. Schizophrenic patients and control subjects did not differ significantly with regard to age. Since racial differences concerning certain metals in the cerebrospinal fluid have been reported (Potkin et al., 1982), only brain material from white subjects was analysed. Histopathological examination was performed on all brains in order to rule out abnormalities such as tumor, infarction, anoxia, brain atrophy and Alzheimer's disease. Post-mortem delay time, i.e. the time between death and the freezing of the brains, was less than 24 hours in all cases. Brain regions investigated included the cortex, gyrus cinguli, caudate nucleus, hippocampus, amygdala, corpus mamillare and hypothalamus.

Following autopsy the brain tissue was quickly frozen and stored at –70 °C until further analysis. Iron content was determined by an atomic absorption procedure (Stevens, 1970). Tissue was freeze-dried at –60 °C and 10^{-2} Torr for 24 hours. Dry tissue was then weighed and dissolved in acid-washed vials with 1 ml of 65% nitric acid at 110 °C. The evaporated dry residue was taken up into 5 ml of the diluent. Atomic absorption spectroscopy was performed using a Zeiss PMQ II photometer adapted for atomic absorption. Results were expressed as µg/g dry weight. Patient and control data were compared using Wilcoxon's rank-sum test (Wilcoxon, 1945).

Results

Analysis of iron content in the brain regions examined showed a clear regional difference (Kruskal-Wallis one-way analysis of variance) with iron levels being highest in the caudate nucleus. Iron content in the schizophrenic group was not different from controls for the cortex, gyrus cinguli, caudate nucleus, hippocampus, amygdala, corpus mamillare and hypothalamus (see Fig. 1). In one female patient who had suffered from tardive dyskinesia prior to death, the iron concentration in all the brain regions examined was within the mean ± 2 S.D. range of those schizophrenic subjects without prior tardive dyskinesia. No correlation between iron content and the neuroleptic-free period prior to death could be found (Spearman's rank correlation).

Discussion

Both the iron concentrations and the distribution of iron within the brain with highest levels in the caudate nucleus are well in accord with previous studies (Hallgren and Sourander, 1958; Harrison et al., 1968; Riederer et al., 1989). The present results showed no difference in post-mortem iron content between schizophrenic patients and control subjects in the cortex, gyrus cinguli, caudate nucleus, hippocampus, amygdala,

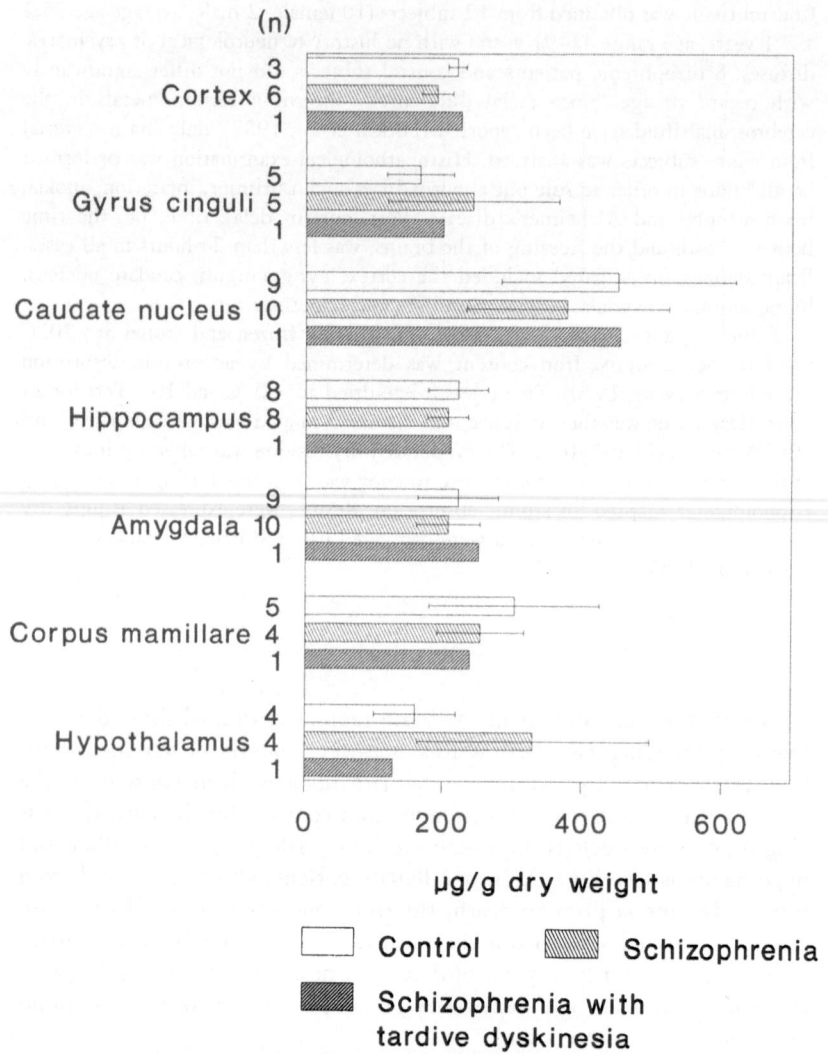

Fig. 1. Iron content as measured by atomic absorption spectroscopy in various brain regions of schizophrenic patients and control subjects. Mean values ± S.D. are presented

corpus mamillare and hypothalamus. By contrast, an increase in iron in the caudate nucleus has been reported using optical density measurements of coronal sections stained with the Perl's technique (Casanova et al., 1992). The authors interpreted their finding as the result of neuroleptic treatment, since the lowest caudate iron-staining in the schizophrenic group was observed in a patient who had not received neuroleptic medication for two years prior to death while the one control subject with abnormally increased iron-staining had been treated with neuroleptics because of drug addiction.

There is indirect evidence of an abnormal iron metabolism in tardive dyskinesia (Weiner et al., 1977; Yehuda and Youdim, 1988). In a female patient with tardive dyskinesia, high, but still normal, iron concentrations have been reported in the substantia nigra and globus pallidus (Hunter et al., 1968). High iron staining has been found in the substantia nigra pars reticulata, putamen, caudate nucleus, globus pallidus and subthalamic nucleus of a male patient who had suffered from haloperidol-induced tardive dyskinesia prior to death (Campbell et al., 1985). Furthermore, a magnetic resonance imaging study of patients with tardive dyskinesia showed shortened T_2 relaxation times in the left caudate nucleus, which may reflect an increased iron content in this brain region (Bartzokis et al., 1990). The single case of tardive dyskinesia reported in the present study did not show a marked difference in iron content from the other schizophrenic patients who had not developed tardive dyskinesia. However, the subcellular distribution of iron could be disturbed in tardive dyskinesia and this may escape detection when, as was the case in the present study, freeze-dried tissue powder is analysed.

The available results concerning iron content in the brain of schizophrenic patients are inconclusive. Early studies performed prior to the introduction of neuroleptic therapy indicate an increase in brain iron in schizophrenia (Josephy, 1930; Hopf, 1952). Unfortunately these reports provided no quantitative data and were based on the subjective impression of the examiners. The results of later post-mortem studies (Casanova et al., 1990, 1992), including the present report, which have attempted to quantify iron content are rendered uncertain by the use of neuroleptic medication. It is theoretically possible that in the latter studies a pre-existing reduction in brain iron in schizophrenic patients was masked by neuroleptics. Further quantitative studies on neuroleptic-free patients with schizophrenia will be needed to clarify this issue.

References

Bartzokis G, Garber HJ, Marder SR, Olendorf WH (1990) MRI in tardive dyskinesia: shortened left caudate T_2. Biol Psychiatry 28: 1027–1036

Campbell WG, Raskind MA, Gordon T, Shaw CM (1985) Iron pigment in the brain of a man with tardive dyskinesia. Am J Psychiatry 142: 364–365

Casanova MF, Waldman IN, Kleinman JE (1990) A postmortem quantitative study of iron in the globus pallidus of schizophrenic patients. Biol Psychiatry 27: 143–149

Casanova MF, Comparini SO, Kim RW, Kleinman JE (1992) Staining intensity of brain iron in patients with schizophrenia: a postmortem study. J Neuropsychiatr Clin Neurosci 4: 36–41

Feighner JP, Robins E, Guze SB, Woodruff RA, Winokur G, Munoz R (1972) Diagnostic criteria for use in psychiatric research. Arch Gen Psychiatry 26: 57–63

Hallgren B, Sourander P (1958) The effect of age on the non-haemin iron in the human brain. J Neurochem 3: 41–51

Harrison WW, Netsky MG, Brown MD (1968) Trace elements in human brain: copper, zinc, iron, and magnesium. Clin Chim Acta 21: 55–60

Hill JM (1988) The distribution of iron in the brain. In: Youdim MBH (ed) Brain iron: neurochemistry and behavioural aspects. Taylor and Francis, London, pp 1–24

Hill JM, Switzer RC (1984) The regional distribution and cellular localization of iron in the rat brain. Neuroscience 11: 595–603

Hopf A (1952) Über histopathologische Veränderungen im Pallidum und Striatum bei Schizophrenie. In: First International Congress of Neuropathology, vol 3. Rosenberg and Sellier, Turin, pp 629–635

Hunter R, Blackwood W, Smith MC, Cumings JN (1968) Neuropathological findings in three cases of persistent dyskinesia following phenothiazine medication. J Neurol Sci 7: 263–273

Josephy H (1930) Dementia praecox (Schizophrenie). In: Bumke O (Hrsg) Die Anatomie der Psychosen. Springer, Berlin

Kornhuber J, Lange KW, Kruzik P, Jellinger K, Gabriel E, Riederer P (1993) The contents of iron, copper, zinc, magnesium, and calcium in post-mortem brain tissue from schizophrenic patients. Biol Psychiatry (submitted)

Lange KW, Youdim MBH, Riederer P (1992) Neurotoxicity and neuroprotection in Parkinson's disease. J Neural Transm [Suppl] 38: 27–44

Löwenthal A, Bruyn GW (1968) Calcification of the striopallidodentate system. In: Vinken PJ, Bruyn GW (eds) Handbook of clinical neurology, vol 6. North-Holland, Amsterdam, pp 703–725

Pollitt E, Leibel R (eds) (1982) Iron deficiency: brain biochemistry and behaviour. Raven Press, New York

Potkin SG, Shore D, Torrey EF, Weinberger DR, Gillin JC, Henkin RI, Agarwal RP, Wyatt RJ (1982) Cerebrospinal fluid zinc concentrations in ex-heroin addicts and patients with schizophrenia: some preliminary observations. Biol Psychiatry 17: 1315–1322

Riederer P, Sofic E, Rausch WD, Schmidt B, Reynolds GP, Jellinger K, Youdim MBH (1989) Transition metals, ferritin, glutathione, and ascorbic acid in parkinsonian brains. J Neurochem 52: 515–520

Spatz H (1924) Über den Eisennachweis im Gehirn, besonders in Zentren des extrapyramidal-motorischen Systems. Z Neurol Psychiat LXXVII: 261–390

Schiffer D (1971) Calcification in the nervous tissue. In: Minkler J (ed) Pathology of the nervous system, vol 3. McGraw-Hill, New York, pp 1342–1360

Smeyers-Verbeke J, Michotte Y, Pelsmaeckers J, et al (1975) The chemical composition of idiopathic nonarteriosclerotic cerebral calcifications. Neurology 25: 48–57

Stevens BJ (1970) Clinical applications of atomic absorption spectroscopy. Varian Techtron Pty Ltd, Australia

Stevens JR (1982) Neuropathology of schizophrenia. Arch Gen Psychiatry 39: 1131–1139

Weiner WJ, Nausieda PA, Klawans HL (1977) Effect of chlorpromazine on central nervous system concentrations of manganese, iron, and copper. Life Sci 20: 1181–1186

Wilcoxon F (1945) Individual comparisons by ranking methods. Biometrics 1: 80–83

Yehuda S, Youdim MBH (1988) Brain iron deficiency: biochemistry and behaviour. In: Youdim MBH (ed) Brain iron: neurochemical and behavioural aspects. Taylor and Francis, London, pp 89–114

Youdim MBH (1988) Brain iron: neurochemical and behavioural aspects. Taylor and Francis, London

Youdim MBH (1990) Developmental neuropharmacological and biochemical aspects of iron-deficiency. In: Dobbing J (ed) Brain, behaviour and iron-deficiency. Springer, Berlin Heidelberg New York Tokyo

Youdim MBH, Ben-Shachar D, Riederer P (1989) Is Parkinson's disease a progressive siderosis of substantia nigra resulting in iron and melanin induced neurodegeneration? Acta Neurol Scand 26: 47–54

Correspondence: Priv.-Doz. Dr. K.W. Lange, Department of Psychiatry, University of Würzburg, Füchsleinstrasse 15, D-97080 Würzburg, Federal Republic of Germany.

Some reflections on iron dependent free radical damage in the central nervous system

M. E. Götz[1], A. Dirr[1], W. Gsell[1], R. Burger[1], A. Freyberger[2], and P. Riederer[1]

[1] Division of Clinical Neurochemistry, Department of Psychiatry, University of Würzburg, and [2] Institute of Toxicology, Bayer AG Pharma Forschungszentrum, Wuppertal, Federal Republic of Germany

Summary

We give a brief introduction into the chemical relationship between iron and oxidative stress serving two purposes:

Firstly we will mention possible deleterious consequences of iron accumulation in central nervous system (CNS) to biomolecules essential for cell viability. Secondly, we try to discuss some difficulties scientists have to face when interpreting experimental data by which they want to prove or to rule out a role of reactive oxygen species (ROS) in the pathogenesis or progression of human diseases.

Organisms take great care in the handling of iron using both transport and storage proteins to minimize the amount of so called low molecular weight iron within the cells and in extracellular fluids (Halliwell and Gutteridge, 1992; Connor and Benković, 1992). The same is true of copper. This sequestration of transition metals may be regarded as an important contribution to tissue defense against oxidative stress. Disturbance of transition metal homeostasis as it is severely observed in CNS trauma, stroke and inflammatory diseases are sometimes supposed to initiate or at least to contribute to the disease process. In addition, other "injury mediators" are supposed to also be involved such as prostaglandins, leukotrienes, interleukins, interferons and tumor necrosis factor. With respect to iron toxicity two main crucial questions have to be posed:

– What are the chemical mediators of transition metal toxicity?
– Is the formation of these mediators pathologically and clinically relevant?

The presence of iron in the active site of many enzymes including those of complexes of electron transfer chains underline its high reactivity towards oxygen species thereby accelerating oxidation of a huge variety of biomolecules (Youdim, 1988).

Iron and copper are effective in converting less reactive species to more reactive ones for example by oxidation of ascorbic acid (to semide-hydroascorbic acid), thiols, catecholamines or hydrogen peroxide. The latter, hydrogen peroxide plays a key role in the formation of cytotoxic hydroxyl radical (OH)˙ via the Fenton reaction involving ferrous iron (Shen et al., 1992; Croft et al., 1992) (Fig. 1).

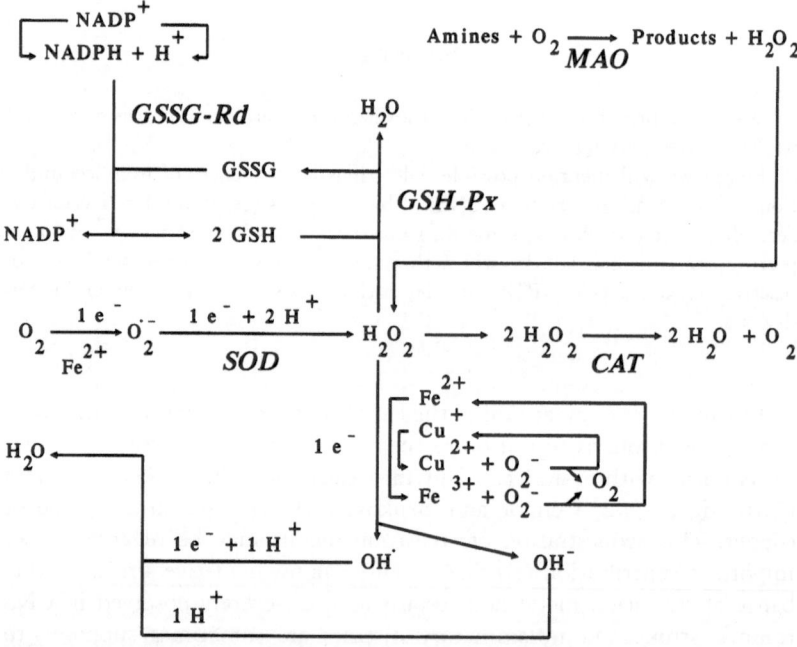

Fig. 1. Possible redox reactions leading to reactive oxygen species (hydrogen peroxide, H_2O_2; superoxide $(O_2)^{˙-}$; hydroxyl radical, $(OH)^{˙}$) and pathways degradating hydrogen peroxide directly via catalase or via NADPH dependent mechanisms utilizing glutathione (*GSH*). *SOD* superoxide dismutase; *GSSG-Rd* glutathione disulfide reductase; *GSH-Px* glutathione peroxidase; *CAT* catalase; *MAO* monoamine oxidases A and B (adapted from Benzi et al., 1988)

In brain hydrogen peroxide is enzymatically generated mainly from dismutation of superoxide, catalyzed by superoxide dismutases (SOD) (Fridovich, 1975, 1986a, b).

Especially in biogenic amine rich brain regions a major source of hydrogen peroxide is the deamination of catechol- and indolamines by flavin containing monoamine oxidases A and B (for recent reviews see: Youdim and Finberg, 1990; Gerlach et al., 1992).

The major source of superoxide in a healthy cell is the activity of electron transport chains in mitochondria and in endoplasmic reticulum (Gorsky et al., 1984).

Some of the electrons passing through these chains "leak" directly from intermediate electron carriers (possibly iron complexes) onto dioxygen. The rate of leakage at physiological concentrations is estimated to be not more than 5% of total electron flow (Boveris and Chance, 1973; Paraidathathu et al., 1992; Sohal and Brunk, 1992).

Aldehyde oxidase and xanthine oxidase are other sources generating superoxide. Xanthine oxidase can be converted from a non-superoxide producing dehydrogenase to the oxidase form during tissue hypoxia possibly by calcium dependent proteases (Granger et al., 1981). Thus, this pathway seems to become an important source of superoxide especially during ischemia-reperfusion injury (Ikeda and Long, 1990).

Hence the toxicity of excess of dioxygen may be due to increased formation of superoxide which is reported to accelerate autoxidations of, for example, catecholamines. However it is also hypothesized that superoxide has first to be dismutated by SOD to hydrogen peroxide to provide the more reactive and more toxic hydroxyl radical (Fig. 1).

Besides the formation of hydroxyl radical other highly reactive oxygen-iron complexes such as perferryl ($Fe^{2+} - O_2^-$) or ferryl iron (Fe = O^{2+}) with iron bearing oxidation numbers of III and IV, respectively are considered to be involved in deleterious reactions with biomolecules (Shen et al., 1992).

Thus, the formation of ROS seems likely if low molecular weight iron species not bound to storage or transport proteins exist in the µMolar range in a cell compartment such as the mitochondria.

To judge the relevance of iron for tissue injury one has to remember the existence of a great variety of defense mechanisms against ROS, their reaction products such as lipid and water soluble antioxidants and antioxidative enzymes such as catalase (CAT) in microperoxysomes and glutathione peroxidase in mitochondria, microsomes and cytosol, both enzymes degrading hydrogen peroxide to water (Fig. 2).

If these and other repair systems fail (Rao and Loeb, 1992), the consequence will be the oxidation of proteins, nucleic acids or lipids depending on the compartment in which ROS occur in excess.

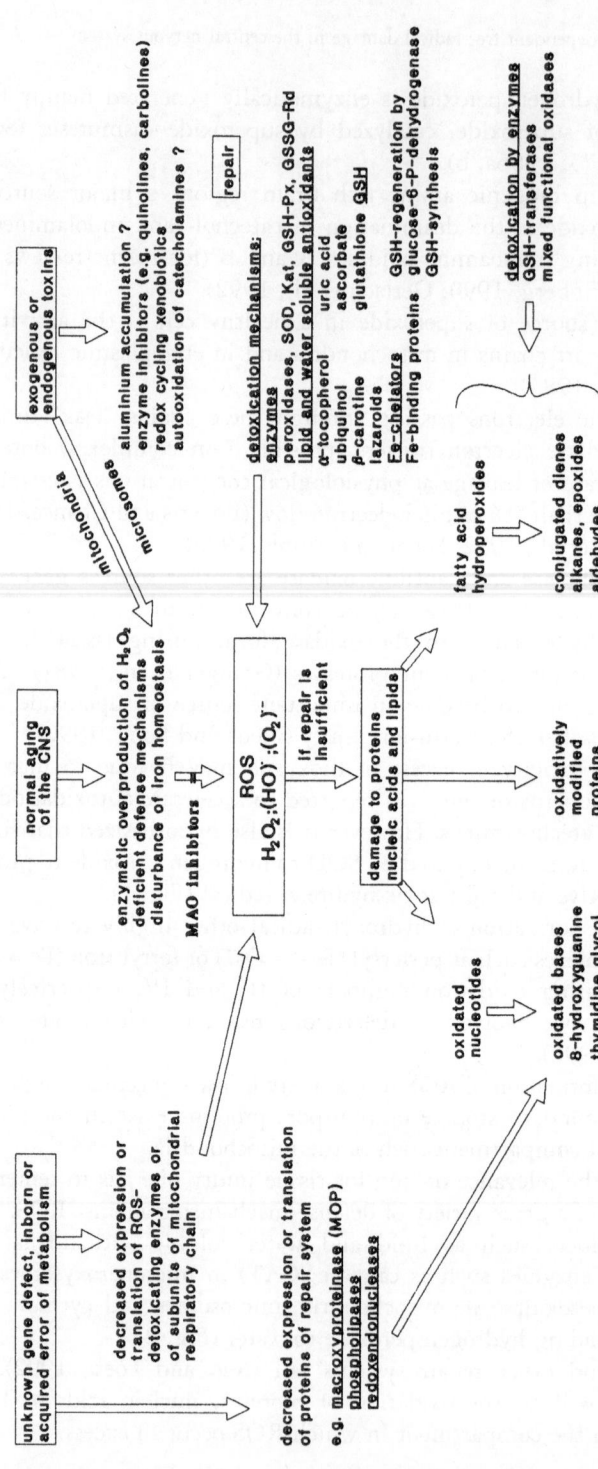

Fig. 2. Schematic representation of putative pathogenetic causes assumed to contribute to ROS toxicity, consequently leading to damage to biomolecules if repair mechanisms become insufficient. Putative sites of therapeutic intervention is indicated by double line crossing arrows

Several criteria should ideally be fulfilled to prove that ROS are important in a particular CNS disease (Halliwell, 1992a, b).

Firstly, the existence of ROS at the injury site has to be demonstrated and secondly, the time course of their formation should correlate with disease pathology.

Locally increased levels of reactive iron and hydrogen peroxide at the same place and time could impose a constant threat to cellular defense mechanisms. If defense and repair are overcome cell damage will occur. This can be a slow process which has been hypothetically correlated to the gradual progression of neurodegenerative diseases (Fig. 2).

Thus, impairment of cell function because of aging, toxins, inborn or acquired genetic defects could lead to a disbalance of damage and repair favouring damage.

Thirdly, the removal of ROS or prevention of their formation should have beneficial effects to pathology and clinic.

This is difficult to achieve. Systemic administered drugs or their active metabolites always have to pass the blood brain barrier. Although most lipid soluble antioxidants enter the brain, they are only effective in antagonizing *lipid* peroxidation. In contrast, events occurring in the cytoplasm are likely to be unaffected by lipophilic antioxidants. The DATATOP study (Deprenyl and Tocopherol antioxidant therapy of Parkinsonism study) using selegiline (monoamine oxidase B inhibitor) and alpha-tocopherol (lipid soluble antioxidant) as antioxidant therapy of Parkinson's disease did not reveal beneficial effects of tocopherol (The Parkinson Study Group, 1993). However, we would only be able to dismiss the hypothesis that ROS participate in neurodegeneration if there were water soluble antioxidants that could not stop the progress of the disease although passing the blood brain barrier and being active at the affected brain areas. This has not yet been proven.

Fourthly, direct application of ROS found in vivo should ideally reproduce most or all pathological and clinical signs of the injury or disease.

For many of the currently used drugs to induce brain damage in vivo the involvement of ROS in their pathological mode of action is not yet proven. It is supposed that ROS are involved in the action of redox cycling compounds such as paraquat or antitumor quinones (Powis, 1989; Sinha and Mimnaugh, 1990) or compounds which readily autoxidize such as 6-hydroxydopamine especially in the presence of ferric iron (Kostrzewa, 1989).

All these criteria are subject to intensive research. Especially the search for ROS or its reaction products is becoming highly sophisticated due to the usage of modern analytical techniques formerly not at the disposal of scientists.

Direct measurement of ROS requires very rapid detection methods in the time range of micro to milliseconds (for example by pulse radiolysis and electron paramagnetic resonance). Indirect proof for radical participation of ROS in an in vivo process necessarily lacks specificity, since the radicals may have reacted with different biomolecules depending on the local circumstances within the cell (Saran and Bors, 1991).

The reaction products of ROS include oxidized proteins, hydroxylated nucleosides or degradation products of peroxides of polyunsaturated fatty acids, namely volatile alkanes, alpha-beta unsaturated aldehydes and malonaldehyde commonly known as malondialdehyde (Pryor and Godber, 1991; Gutteridge and Halliwell, 1990).

Recently analytical chemistry using gas chromatography-mass spectrometry reports success in the detection of small amounts of adducts of products of fatty acid peroxidation and proteins or nucleic acids (Hageman et al., 1992).

For a long time however, the most widely used assay has been the measurement of aldehydic degradation products of fatty acid peroxides by the thiobarbituric acid test (Janero, 1990; Valenzuela, 1991). The lipid material is simply heated with thiobarbituric acid (TBA) at low proton concentration and the formation of a pink chromogen is measured at 532 nm. In addition to malonaldehyde other aldehydes yield chromogens with TBA having similar absorption maxima. This is the reason why the term thiobarbituric acid reactive substances (TBARS) is preferred.

Although simple, rapid and convenient in comparing a large number of samples simultaneously the assay is highly dependent on many standardized reaction conditions including degree of polyunsaturated fatty acid unsaturation, proton donors, antioxidants, temperature, pH, solvent, oxygen concentration and transition metal content.

In order to evaluate the extent of chromogen formation which is produced by in vivo *preexisting* lipid peroxides and that which is *newly* produced during the assay procedure we investigated the influence of oxygen and iron on test response.

The presence of oxygen clearly influences the amount of pink pigment formed during the acid heating step of the assay being true for lipid extracts as well as for homogenates although there had not been any preincubation. Therefore all, the amount of oxygen in the reaction vial, the amount of antioxidants and that of oxidizable substrate determine the rate of autoxidation of polyunsaturated fatty acids (Götz et al., 1993). When different amounts of lipid or homogenate were preincubated under air for 90 minutes at 37 °C enzymatical and chemical oxidation occur being linear in a certain range of substrate as indicated in this figure. However the same amounts of substrate investigated under argon resulted in much lower values of the pink pigment (Fig. 3).

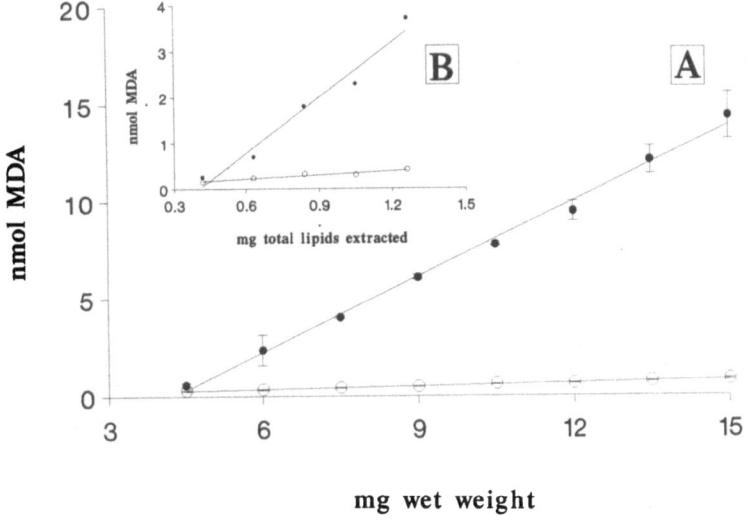

mg wet weight

Fig. 3. Effect of argon (○) or air (●) on the production of TBARS from human cortical homogenates (from 4.5 to 15 mg wet weight (ww). (**A**) and of lipids extracted from the same pool of human cortex (0.4–1.2 mg dry weight. (**B**, insert) preincubated at 37 °C for 90 min in a total volume of 400 μl and subsequently incubated at 95 °C for 75 min in a total volume of 2 ml. Concerning homogenates, values are means of two determinations and expressed as nmol MDA equivalents per assay [from Götz, et al (1993) Neurochem Int 22: 255–262]

The effects of Fe^{3+} and Fe^{2+} on air – stimulated TBARS formation are shown in Fig. 4. In a preincubation step at pH 7,4 the formation of TBARS is enhanced by addition of both Fe^{2+} and Fe^{3+} whilst without preincubation at pH 3,5 Fe^{2+} is far less potent than Fe^{3+} in stimulating lipid peroxidation. For efficient induction of formation of TBARS in vitro in brain tissue homogenates a conversion of Fe^{2+} to Fe^{3+} favoured by neutral or alkaline pH seems to be necessary.

These results indicate, that even only the heating step of the thiobarbituric acid test increases the formation of chromogen nearly ten fold. If an additional preincubation step is performed in the presence of iron even higher values are obtained. Therefore one has to ask for a real basal value indicating the amounts of in vivo preformed peroxidized polyunsaturated fatty acids of biological tissue homogenates.

Thus, the thiobarbituric acid test can serve as a screening method for measuring the *susceptibility* of homogenates or lipid extracts to oxygenation when performed aerobically. If it is kept in mind, that this susceptibility depends mainly on antioxidant status, iron content as well as on the amount of unsaturated fatty acids the thiobarbituric acid test is likely

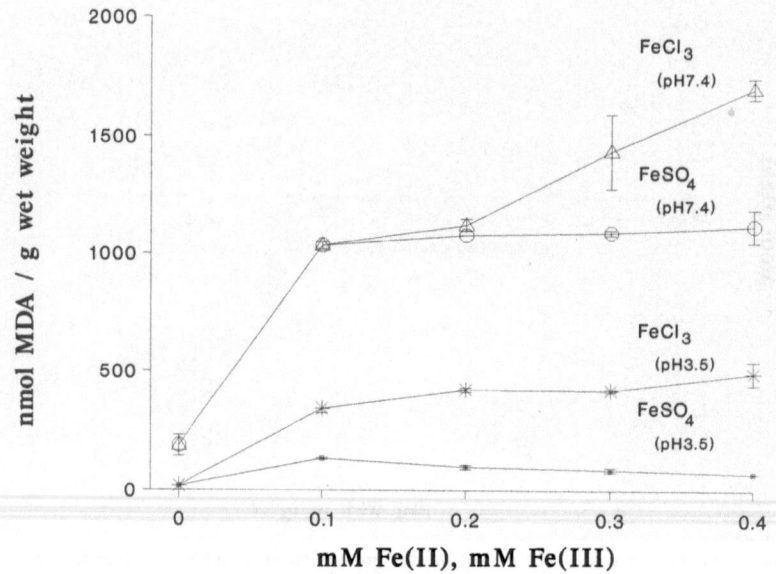

Fig. 4. Human cortical homogenates (60 µl 5% wt/vol 154 mM sodium chloride) were only incubated under air at *pH 3.5* at 95 °C for 75 min without, and in presence of FeSO₄ (100–400 µM) (×), or FeCl₃ (100–400 µM) (○) in a total volume of 2 ml. Next to a preincubation at *pH 7.4* at 37 °C for 90 min in a volume of 400 µl, human cortical homogenates (60 µl 5% wt/vol 154 mM sodium chloride) were incubated at 95 °C for 75 min under air without, and in presence of FeSO₄ (100–400 µM) (△), or FeCl₃ (100–400 µM) (●) in a total volume of 2 ml. Values are means of two determinations and expressed as nmol MDA equivalents/g wet weight (ww) [from Götz, et al (1993) Neurochem Int 22: 255–262]

to give a first impression of the balance between these parameters within the tissue investigated.

Only if driven anaerobically with pure lipids not contaminated with antioxidants, the test is likely to approximate basal levels of lipophilic peroxides. Thus, the exact reaction conditions to be used depend on the system investigated (i.e. homogenates or lipid extracts) and the conclusions which are to be drawn by means of this assay.

However, measurements of TBARS not sufficiently evidence the involvement of oxidative stress and lipid peroxidation in pathogenetic processes albeit often stated in the literature.

In conclusion, in order to evaluate the importance of oxidative stress for pathogenetic processes, other possible markers such as levels of antioxidants, unsaturated fatty acids and transition metals should be correlated with data obtained by measurements of TBARS. This example demonstrates only one of many difficulties scientists have to face when quantitation of the real occurrence of ROS and iron toxicity in vivo becomes necessary.

Acknowledgements

We wish to thank Dr. Y. Taneli for helpful discussions. This work was supported by a grant from the Bundesministerium für Forschung und Technologie (BMFT) grant number 01 KL 9101-0.

References

Benzi G, Pastoris O, Villa RF (1988) Changes induced by aging and drug treatment on cerebral enzymatic antioxidant system. Neurochem Res 13: 467–478

Boveris A, Chance B (1973) The mitochondrial generation of hydrogen peroxide. General properties and effect of hyperbaric oxygen. Biochem J 134: 707–716

Connor JR, Benković SA (1992) Iron regulation in the brain: histochemical, biochemical, and molecular considerations. Ann Neurol 32: S51–S61

Croft S, Gilbert BC, Lindsay Smith JR, Whitwood AC (1992) An E.S.R. investigation of the reactive intermediate generated in the reaction between FeII and H_2O_2 in aqueous solution of the hydroxyl radical. Free Radic Res Commun 17: 21–39

Fridovich I (1975) Superoxide dismutases. Annu Rev Biochem 44: 147–159

Fridovich I (1986a) Biological effects of the superoxide radical. Arch Biochem Biophys 247: 1–11

Fridovich I (1986b) Superoxide dismutases. Adv Enzymol 58: 62–97

Gerlach M, Riederer P, Youdim MBH (1992) The molecular pharmacology of L-deprenyl. Eur J Pharmacol-Mol Pharmacol Sect 226: 97–108

Götz ME, Dirr A, Freyberger A, Burger R, Riederer P (1993) The thiobarbituric acid assay reflects susceptibility to oxygen-induced lipid peroxidation in vitro rather than levels of lipid hydroperoxides in vivo: a methodological approach. Neurochem Int 22: 255–262

Gorsky LD, Koop DR, Coon MJ (1984) On the stoichiometry of the oxidase and monooxygenase reactions catalyzed by liver microsomal cytochrome P-450. J Biol Chem 259: 6812–6817

Granger DN, Rutili G, McCord JM (1981) Superoxide radicals in feline intestinal ischemia. Gastroenterology 81: 22–29

Gutteridge JMC, Halliwell B (1990) The measurement and mechanism of lipid peroxidation in biological systems. Trends Biol Sci 4: 129–135

Hageman JJ, Bast A, Vermeulen NPE (1992) Monitoring of oxidative free radical damage in vivo: analytical aspects. Chem-Biol Interactions 82: 243–293

Halliwell B (1992a) Oxygen radicals as key mediators in neurological disease: fact or fiction? Ann Neurol 32: S10–S15

Halliwell B (1992b) Reactive oxygen species and the central nervous system. J Neurochem 59: 1609–1623

Halliwell B, Gutteridge MC (1992) Biologically relevant metal ion-dependent hydroxyl radical generation: an update. FEBS Lett 307: 108–112

Ikeda Y, Long DM (1990) The molecular basis of brain injury and brain edema: the role of oxygen free radicals. Neurosurgery 27: 1–11

Janero DR (1990) Malondialdehyde and thiobarbituric acid-reactivity as diagnostic indices of lipid peroxidation and peroxidative tissue injury. Free Radic Biol Med 9: 515–540

Kostrzewa RM (1989) Neurotoxins that affect central and peripheral catecholamine neurons, In: Boulton AB, Baker GB, Juorio AV (eds) Neuromethods, vol 12. Drugs as tools in neurotransmitter research. Humana Press, Clifton, p 1

Paraidathathu T, De Groot H, Kehrer JP (1992) Production of reactive oxygen by mitochondria from normoxic and hypoxic rat heart tissue. Free Radic Biol Med 13: 289–297

Powis G (1989) Free radical formation by antitumor quinones. Free Radic Biol Med 6: 63–101

Pryor WA, Godber SS (1991) Noninvasive measures of oxidative stress status in humans. Free Radic Biol Med 10: 177–184

Rao KS, Loeb LA (1992) DNA damage and repair in brain: relationship to aging. Mutat Res 275: 317–329

Saran M, Bors W (1991) Direct and indirect measurements of oxygen radicals. Clin Invest 69: 957–964

Shen X, Tian J, Li Y, Li X, Chen Y (1992) Formation of the excited ferryl species following Fenton reaction. Free Radic Biol Med 13: 585–592

Sinha BK, Mimnaugh EG (1990) Free radicals and anticancer drug resistance: oxygen free radicals in the mechanisms of drug cytotoxicity and resistance by certain tumors. Free Radic Biol Med 8: 567–581

Sohal RS, Brunk UT (1992) Mitochondrial production of pro-oxidants and cellular senescence. Mutat Res 275: 295–304

The Parkinson Study Group (1993) Effects of tocopherol and deprenyl on the progression of disability in early Parkinson's disease. N Engl J Med 328: 176–183

Valenzuela A (1991) The biological significance of malondialdehyde determination in the assessment of tissue oxidative stress. Life Sci 48: 301–309

Youdim MBH (1988) Brain iron: neurochemical and behavioural aspects. Taylor & Francis, New York, p 1

Youdim MBH, Finberg JPM (1990) New directions in monoamine oxidase A and B: selective inhibitors and substrates. Biochem Pharmacol 41: 155–162

Correspondence: Dr. M.E. Götz, Department of Psychiatry, Division of Clinical Neurochemistry, University of Würzburg, Füchsleinstrasse 15, D-97080 Würzburg, Federal Republic of Germany.

Iron regulation of dopaminergic transmission: relevance to movement disorders

D. Ben-Shachar, A. Tovi, and M. B. H. Youdim

Department of Pharmacology, Faculty of Medicine, Technion, Haifa, Israel

Summary

Dopamine has been implicated in many central and peripheral functions, among which are motor control, cognitive, affective and neuroendocrine functions. We suggest that iron plays a crucial role in the dopamine system, being essential for the normal functioning of several components in the metabolism and transmission of dopamine. In 6-hydroxydopamine treated rats, which serve as an animal model of Parkinson's disease, chelation of iron by desferrioxamine retarded the degeneration, expressed both biochemically and behaviorally, of the nigrostriatal dopamine neurons. Moreover, introducing iron into the substantia nigra of the rat induced parkinsonian-like biochemical and behavioral responses. Long term neuroleptic treatment in schizophrenia, which is associated with extrapyramidal motor side effects, the most severe of which is tardive dyskinesia, caused in rats dopamine supersensitivity, which could be regulated by iron supplementation or deprivation. This was accompanied by a significant transport of iron into the brain which is normally very limited. In addition there was a definite correlation between the ability of the neuroleptic drug to induce iron transport into the brain and its ability to induce extrapyramidal severe motor side effects. The significance of brain iron homeostasis to dopamine normal motor and biochemical function in animal models is argued.

Introduction

Abnormality of the central dopaminergic system is often associated with movement disorders as in Parkinson's disease and tardive dyskinesia, or with mental defects as in schizophrenia. Iron plays a crucial role in the regulation of dopamine function (Ben-Shachar et al., 1985, 1986; Youdim et al., 1989; Ben-Shachar and Youdim, 1990). This is not totally

D. Ben-Shachar et al.

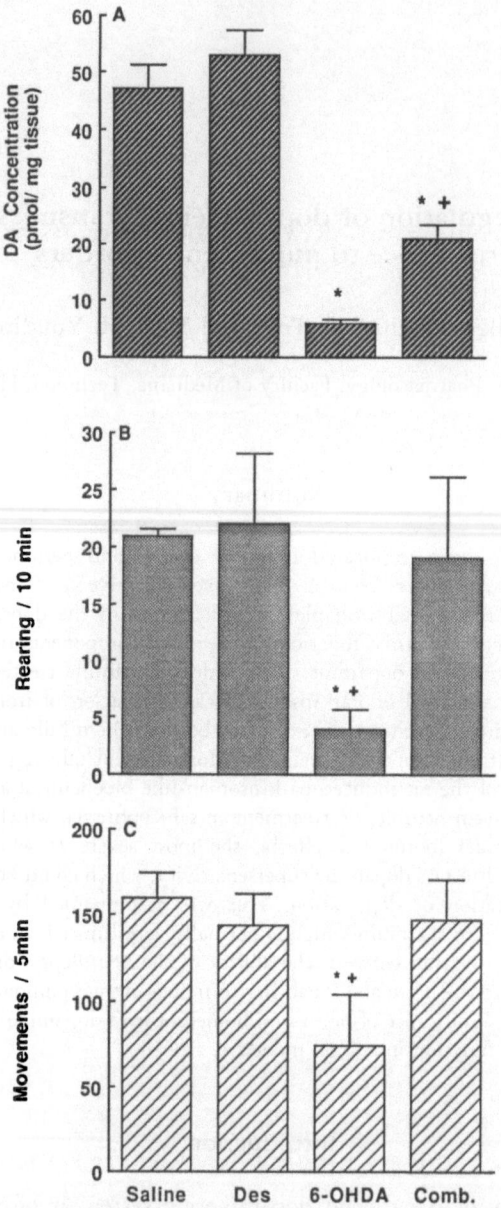

Fig. 1. Striatal dopamine concentration (**A**), rearing activity (**B**) and spontaneous movements in a novel space (**C**) measured in rats which were injected intraventricularly with saline, desferrioxamine (*Des*-130 ng), 6-hydroxydopamine (*6-OHDA* 250 µg), or a combination of 6-OHDA and Des (*Comb.*). Adapted from Ben-Shachar et al. (1991b). * p < 0.01 6-OHDA vs. saline. + p < 0.02 < 6-OHDA vs. Comb.

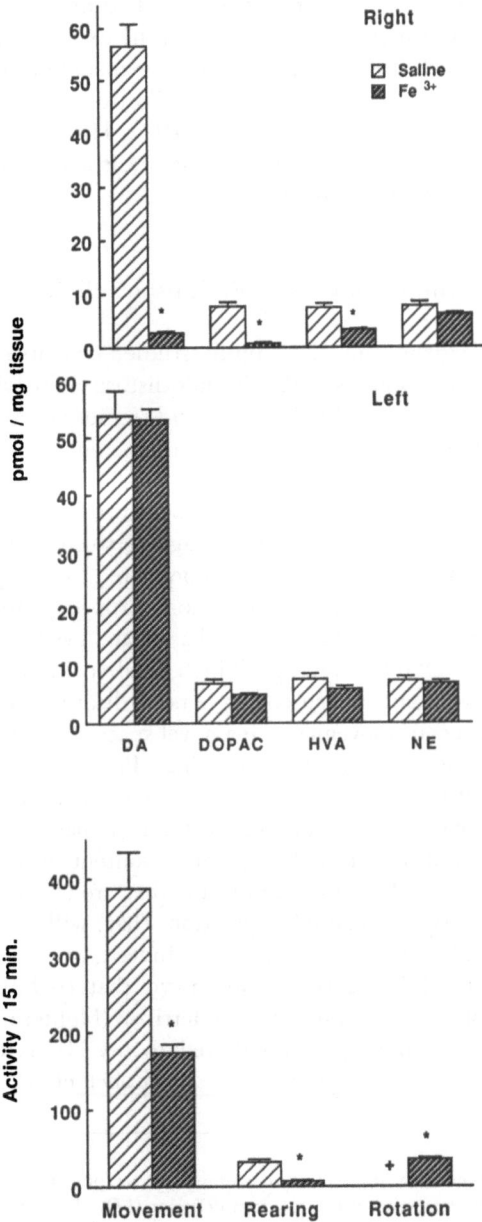

Fig. 2. Dopamine and its metabolites and norepinephrine concentrations in the left and right striata of rats which were injected with iron or saline into the right striatum, as well as spontaneous movements, rearing activity and amphetamine induced ipsilateral rotation. Adapted from Ben-Shachar and Youdim (1991). * $p < 0.001$ + no rotation

unexpected since iron is a cofactor of tyrosine hydroxylase (Nagatsu et al., 1964) and probably participates in monoamine oxidase synthesis (Symes et al., 1969), which are key enzymes in the metabolism of dopamine. Moreover, the unique pattern of the distribution of iron in the brain almost parallels that of dopamine (Spatz, 1922; Hill and Switzer, 1984). The linkage between a defect in iron homeostasis and dopamine dependent movement disorders is discussed.

Animal models of Parkinson's disease

The most evident and thoroughly studied example of dopamine derived movement disorders is Parkinson's disease (PD), which is characterized mainly by akinesia, bradykinesia and tremor at rest. Post mortem studies of parkinsonian brains revealed that iron accumulates in the substantial nigra pars compacta (Sofic et al., 1988; Dexter et al., 1989; Perl et al., 1993), the same region which shows the highest degree of dopamine neuron degeneration. The linkage between iron and dopamine dependent movement disorders was demonstrated by studying the effect of the iron chelator, desferrioxamine, on 6-hydroxydopamine (6-OHDA) induced parkinsonism in rats (Ben-Shachar et al., 1991b). Intraventricular injection of 6-OHDA (250 μg), which resulted in 90% reduction in striatal dopamine content, caused movement disorders as expressed by decreased spontaneous movements in a novel space and reduced dopamine dependent rearing, sniffing and grooming. Pretreatment of 6-OHDA injected rats with desferrioxamine (130 ng) normalized dopamine dependent behavioral responses in spite of the fact that 6-OHDA-induced lesion as expressed by striatal dopamine content was only partially normalized (Fig. 1). This may implicate that iron is necessary for 6-OHDA-induced severe neurodegeneration that will be expressed in dopamine dependent behavioral responses. Indeed, in-vitro studies on the mechanism of 6-OHDA action have shown that 6-OHDA is able to release iron from its binding sites in ferritin (Monterio and Winterbourne, 1989), which may be attributed to 6-OHDA linkage to free oxygen radical generation and neurotoxicity (Cohen et al., 1974; Graham

$\longrightarrow$

Fig. 3. Caudate non-heme iron levels, dopamine D_2 receptor number and dopamine dependent behavioral response in control, iron deficient (*I.D.*), haloperidol (*Hal*) and iron deficient + haloperidol (*I.D.* + *Hal*) treated rats. *p < 0.005 treatment vs. control; + p < 0.005 treatment vs. ID. The table represents the extent of supersensitivity of the dopamine D2 receptor and apomorphine induced movements as expressed by ratios between the haloperidol treated groups and their respective control groups. * p < 0.05 ** p < 0.001. Adapted from Ben-Shachar et al. (1990)

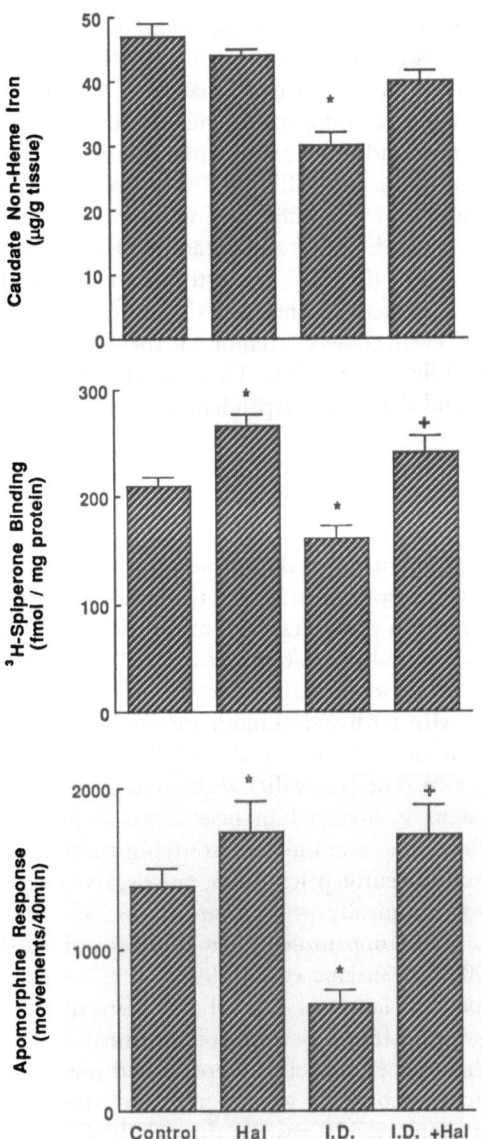

	hal/cont	(ID+hal)/ID
D$_2$ receptor	1.27±0.05	1.49±0.04*
Apomorphine induced response	1.25±0.10	2.3±0.07**

et al., 1978). Unilateral intranigral injection of iron (Fe^{3+} – 50 μg) further supports the suggestion that iron is essential for causing parkinsonian-like symptoms in rat (Ben-Shachar and Youdim, 1991). Four weeks after operation striatal dopamine and its metabolites were severely reduced in the operated side whereas norepinephrine, as well as serotonin and its metabolite, were not affected. This specific damage to the nigrostriatal dopaminergic neurons was accompanied by spontaneous rotations which were amplified by amphetamine (Fig. 2). Moreover, these rats exhibited movement disorders characteristic of PD, such as akinesia which was expressed by reduced horizontal movements and rearing in a novel space. They even showed tremor of the contralateral paw and freezing in the middle of a motion. These results strongly support the link between iron and dopamine-dependent motor functions.

Neuroleptics, dopamine and iron

Another model which involves dopamine dependent movements disorders is long term neuroleptic medication to schizophrenic patients. This treatment often results in extrapyramidal motor side effects, of which the most severe is tardive dyskinesia (Klawans, 1973). It is of interest that several studies reported an increase in iron concentrations in the basal ganglia of patients with tardive dyskinesia measured both by histochemical and MRI techniques (Hunter et al., 1968; Campbell et al., 1985; Bratzokis et al., 1990). The possibility that interaction between iron and the dopamine system is involved in neuroleptic-induced severe extrapyramidal side effects was examined by studying the role of iron in the mechanism of action of neuroleptics. Thus, rats deprived or supplied with iron, were treated chronically with neuroleptics, and biochemical or behavioral aspects of the dopamine system were examined (Ben-Shachar and Youdim, 1990; Ben-Shachar et al., 1991a).

Nutritional iron deficient and control rats were injected daily with haloperidol (5 mg/kg). After 2–3 weeks of treatment, followed by three days of drug washout, dopamine D_2 receptors and dopamine dependent motor responses to apomorphine were significantly increased in control rats. However, this potentiation was significantly higher in iron deficient rats treated chronically with haloperidol as compared with control rats

$\longrightarrow$

Fig. 4. Caudate non-heme iron levels, dopamine dependent behavioral response and dopamine D_2 receptor number in control, iron (Fe^{3+}), chlorpromazine (CPZ), and a combination of iron and CPZ treated rats. Adapted from Ben-Shachar et al. (1991a).
* $p < 0.001$

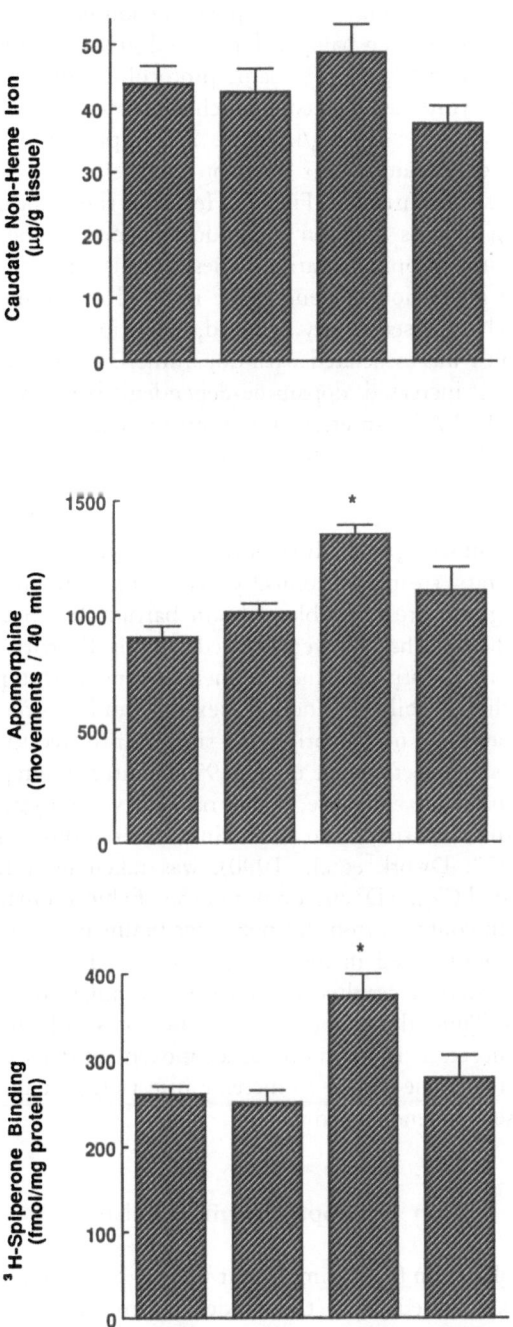

treated with haloperidol (Fig. 3). This phenomenon is further clarified if expressed as a ratio between haloperidol treated groups and their respective controls (Table in Fig. 3). The same protocol of drug treatment was followed, but this time rats treated with chlorpromazine (10 mg/kg/day) were provided with Fe^{3+} (5 mg/kg/day). A complete blockade of the supersensitivity of dopamine D_2 receptor and behavioral responses to apomorphine was demonstrated (Fig. 4). However, the clinical manifestation of the drug such as sedation and reduced rectal temperature, were not affected by iron supplementation. These results suggest that iron is involved in the regulation of neuroleptic-induced dopamine behavioral and biochemical supersensitivity. Indeed, injection of iron into the amygdala complex was associated with dopamine D_2 receptor supersensitivity as well as increased dopamine-dependent behavioral responses (Csernansky et al., 1983). Interestingly, schizophrenia which is characterized by abnormal activation of the dopaminergic system is not associated with typical movement disorders.

The role of iron in neuroleptic mechanism of action was further investigated by following the ability of radioactive iron, injected into the peritoneum of rats treated chronically with three different types of neuroleptic drugs, to cross the blood brain barrier (Ben-Shachar et al., 1993 [unpublished]). The rats were treated chronically with haloperidol – a butyrophenone, chlorpromazine – a phenotyazine, both typical neuroleptics with high probability of inducing extrapyramidal side effects, and clozapine – an atypical neuroleptic with significantly reduced tendency to induce these side effects (Stille et al., 1971). Autoradiography of brain slices from rats one hour after injection of radioactive iron was performed. Iron, whose normal transport into the brain is limited (Fig. 5 B; Dallman and Spirito, 1977; Dwork et al., 1990), was taken up into brains of haloperidol-treated (Fig. 5 D) and even more so of chlorpromazine-treated (Fig. 5 C) rats. In contrast, iron did not enter brains of clozapine treated rats but rather sedimented in blood vessels (Fig. 5 E). Co-treatment of iron and chlorpromazine resulted in massive accumulation of iron in the brain (Fig. 5 F). Thus, dopamine receptor antagonist which are able to change iron homeostasis in brain do induce movement disorders, whereas clozapine which lack the ability to increase iron transport into the brain has a much lesser tendency to do so.

Iron and dopamine metabolism

A partial explanation for the important role of iron in dopamine intact function can be derived from the model of iron deficiency in rats. Nutritional iron deficiency which caused 60% reduction in haemoglobin

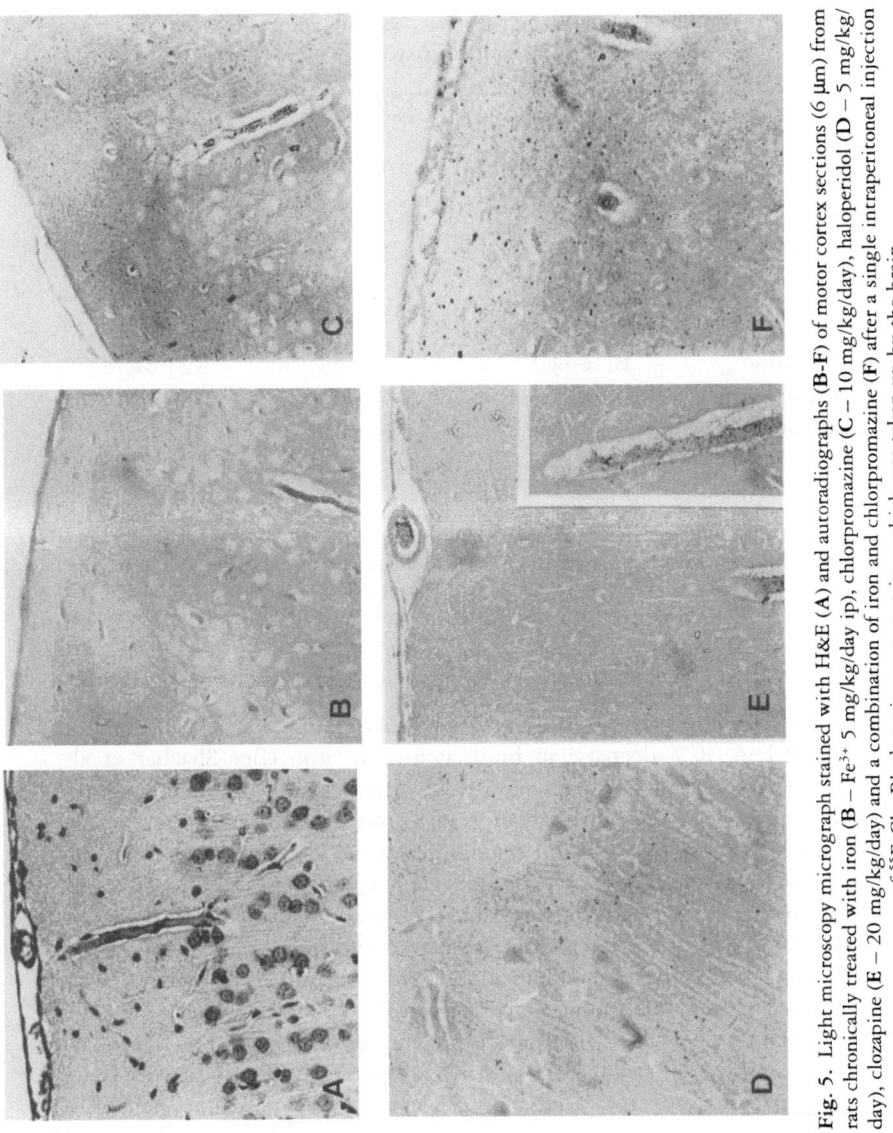

Fig. 5. Light microscopy micrograph stained with H&E (**A**) and autoradiographs (**B-F**) of motor cortex sections (6 μm) from rats chronically treated with iron (**B** – Fe^{3+} 5 mg/kg/day ip), chlorpromazine (**C** – 10 mg/kg/day), haloperidol (**D** – 5 mg/kg/day), clozapine (**E** – 20 mg/kg/day) and a combination of iron and chlorpromazine (**F**) after a single intraperitoneal injection of $^{55}FeCl_3$. Black grains represent iron which was taken up by the brain

Table 1. The effect of iron deficiency on various components of the striatal dopaminergic synapse

		Control	Iron deficiency
DA D_1 receptors	B_{max} (pmol/mg/prot.)	0.95 ± 0.12	0.81 ± 0.09
	K_D (nM)	0.67 ± 0.07	$0.39 \pm 0.04 **$
DA D_2 receptors	B_{max} (pmol/mg/prot.)	0.48 ± 0.04	$0.22 \pm 0.02 ***$
	K_D (nM)	0.74 ± 0.07	0.78 ± 0.05
DA dependent AC activity	D_1 coupled	15.20 ± 1.00	$11.00 \pm 0.70 **$
	D_2 coupled	2.04 ± 0.37	$3.24 \pm 0.15 *$
MAO-B activity	Extraneuronal	2.70 ± 0.25	$1.49 \pm 0.14 **$
DA turnover (DOPAC+HVA)/DA		0.18 ± 0.01	$0.12 \pm 0.01 **$

Iron deficient diet was given to 21 days old male rats for four weeks. Control group received the same diet supplemented with iron. Dopamine receptors maximal number of binding sites (B_{max}) and dissociation constant (K_D), dopamine dependent adenylate cyclase (AC) activity, monoamine oxidase B ($MAO\ B$) activity and dopamine turnover were determined in striatal membrane homogenates. Values are means $\pm$ SEM; * $p < 0.025$, ** $p < 0.01$, *** $p < 0.001$

and serum non-haem iron and 85% reduction in liver non-haem iron resulted in 30% decrease in brain non-haem iron (Ben-Shachar et al., 1986). Reduction in brain non-haem iron specifically damaged the dopaminergic system. Other neurotransmitter systems such as adrenergic, cholinergic, serotonergic, and glutaminergic were unaffected at the receptor level by iron deficiency (Youdim et al., 1989; Tovi, 1992). Moreover, iron deficiency had no influence on the noradrenergic and serotonergic systems at the level of the neurotransmitters and their metabolites content (Tovi, 1992). Several components of the dopaminergic synapse were changed by iron deficiency (summarised in Table 1). Dopamine D_2 receptor maximal number (B_{max}) and the dissociation constant (K_D) of dopamine D_1 receptors were reduced by 50% and 40% respectively. The reduced activity of the receptors was manifested in the second messenger system, since D_1-mediated stimulation and D_2-mediated inhibition of dopamine dependent adenylate cyclase activity were expressed to a lesser extent in iron deficiency. It is interesting to note in this connection that adenylate cyclase activity is inhibited in a dose dependent manner by iron, an effect which is potentiated in the presence of GTP (Moser and Schuster, 1990). Iron deficiency also caused a significant decrease in extraneuronal monoamine oxidase B (MAO B) activity which was associated with small but significant reduction in dopamine turnover. Concomitant with these biochemical changes was the decrease in spontaneous movements as well as dopamine-dependent behavioral responses to apomorphine (Ben-Shachar et al., 1986).

The ability of iron to affect the normal activity of several components of the dopaminergic synapse, making it almost impossible for compensatory mechanisms to overcome the defect, may be a crucial basis for the role of iron in intact dopamine motor functions.

Acknowledgments

This study was supported in part by research grant 4–88–9 from the Israel Institute for Psychobiology – the Charles E. Smith Family Foundation, the Chief Scientist's Office, Ministry of Health, Israel, and by Ciba Geigy, Basle, Switzerland.

References

Bartzokis G, Garber HJ, Marder SR, Oldendorf WH (1990) MRI in tardive dyskinesia shortened left caudate T_2. Biol Psychiatry 28: 1027–1036

Ben-Shachar D, Youdim MBH (1990) Neuroleptic-induced supersensitivity and brain iron: iron deficiency and neuroleptic-induced dopamine D_2 receptor supersensitivity. J Neurochem 54: 1136–1141

Ben-Shachar D, Youdim MBH (1991) Intranigral iron injection induces behavioral and biochemical "parkinsonism" in rats. J Neurochem 57: 2133–2135

Ben-Shachar D, Finberg JPM, Youdim MBH (1985) Effect of iron chelators on dopamine D_2 receptors. J Neurochem 45: 999–1005

Ben-Shachar D, Ashkenazi R, Youdim MBH (1986) Long term consequence of early iron deficiency on dopaminergic neurotransmission. Int J Dev Neurosci 4: 91–98

Ben-Shachar D, Pinhassi B, Youdim MBH (1991a) Prevention of neuroleptic-induced dopamine D_2 receptor supersensitivity by chronic iron salt treatment. Eur J Pharmacol 202: 177–183

Ben-Shachar D, Eshel G, Finberg JPM, Youdim MBH (1991b) The iron chelator desferrioxamine (desferal) retards 6-hydroxydopamine-induced degeneration of nigrostriatal dopamine neurons. J Neurochem 56: 1441–1444

Campbell WG, Raskind MA, Gordon T, Shaw CM (1985) Iron pigment in the brain of man with tardive dyskinesia. Am J Psychiatry 142: 362–364

Cohen G, Heikkila RE, McNamee D (1974) The generation of hydrogen peroxide, superoxide radicals and hydroxy radicals by 6-hydroxydopamine dialuric acid and related cytotoxic agents. J Biol Biochem 249: 2447–2459

Csernansky JC, Holman CA, Bounet KA, Garabowsky K, King R, Hollister LE (1983) Dopaminergic supersensitivity at distant sites following induced epileptic foci. Life Sci 32: 385–390

Dallman PR, Spirito RAJ (1977) Brain iron in the rat: extremely slow turnover in normal rats may explain long lasting effects of early iron deficiency. Nutrition 107: 1075–1081

Dexter DT, Wells FR, Lees AJ, Javoy-Agid F, Agid Y, Jenner P, Marsden CD (1989) Increased nigral iron content and alteration in other metal ions occurring in brain in Parkinson's disease. J Neurochem 52: 1830–1836

Dwork AJ, Lawler G, Zybert PA, Durkin M, Osman M, Willson N, Barakai A (1990) An autoradiographic study of uptake and distribution of iron in the brain of young rat. Brain Res 518: 31–39

Graham D, Tiffany SM, Bell WR, Gutknecht WF (1978) Auto-oxidation versus covalent binding quinones as the mechanism of toxicity of dopamine, 6-hydroxydopamine and related compounds towards C 1300 neuroblastoma cells in vitro. Mol Pharmacol 14: 644–653

Hill JM, Switzer RC (1984) The regional distribution and cellular localization of iron in the rat brain. Neuroscience 11: 595–603

Hunter R, Blackwood W, Smith MCJ (1968) Neuropathological findings in three cases of persistent dyskinesia following phenothiazine medication. J Neurol Sci 7: 763–773

Klawans HL, (1973) The pharmacology of tardive dyskinesia. Am J Psychiatry 130: 82–86

Monterio HP, Winterbourne CC (1989) 6-Hydroxydopamine releases iron from ferritin and promotes ferritin-dependent lipid peroxidation. Biochem Pharmacol 38: 4144–4182

Moser A, Schuster O (1990) Iron inhibits dopamine D_1 receptor coupled adenylate cyclase via G-proteins in caudate nucleus of the rat. Biochem Biophys Res Commun 171: 1372–1377

Nagatsu T, Levitt M, Udenfriend S (1964) Tyrosine hydroxylase, the initial step in norepinephrine biosynthesis. J Biochem Chem 239: 2910–2917

Perl DP, Good PF, Olanow CW (1993) Iron and aluminum accumulate in neuromelanin granules of the substantia nigra pars compacta of idiopathic Parkinson's disease. J Neuropathol (in press)

Sofic E, Riederer P, Heinsen H, Youdim MBH (1988) Increased iron III and total iron content in post mortem substantia nigra of parkinsonian brain. J Neural Transm 74: 199–205

Spatz H (1992) Über den Eisennachweis im Gehirn, besonders in Zentren des extrapyramidal motorischen Systems. Z Ges Neurol Psychiatry 28: 261–390

Stille G, Lauener H, Eichenberger E (1971) The pharmacology of 8-chloro-11-(4-methyl-1-piperazinyl)-5H-dibenzo [b,e] [1,4] diazapine (clozapine). Il Pharmaco 26: 603–625

Symes AL, Sourker TL Youdim MBH, Gregoriadis G, Birnbaum H (1969) Decreased monoamine oxidase activity in liver of iron deficient rats. Can J Biochem 47: 999–1002

Tovi A (1992) The effect of iron deficiency on the metabolism of neurotransmitters in the brain. Thesis, Technion, Haifa, Israel

Youdim MBH, Ben-Shachar D, Yehuda S (1989) Putative biological mechanism of the effect of iron deficiency on brain biochemistry and behaviour. Am J Nutr [Suppl] 50: 607–617

Correspondence: Dr. D. Ben-Shachar, Department of Pharmacology, Faculty of Medicine, Technion, POB 9649, Haifa, Israel.

Dopaminergic cell death in Parkinson's disease: a role of iron?

F. Javoy-Agid and B. Faucheux

INSERM U 289, Hôpital de la Salpêtrière, Paris, France

Summary

Loss of midbrain dopaminergic neurons in Parkinson's disease is associated to hyperoxidation phenomena. In the human substantia nigra, free radicals may be produced in large quantities i.e, during degradation of dopamine, synthesis and accumulation of neuromelanin, through iron, present in high concentrations. Under normal conditions, production of free radicals is compensated by powerful protective enzymes: superoxide dismutase is detected in neurons, and expressed at high levels in those of the substantia nigra; gluthatione peroxidase is exclusively detected in glial cells. A low density of glial cells surround the substantia nigra neurons relatively to other midbrain areas. Thus nigral dopaminergic neurons may be less protected against deleterious action of free radicals. This may explain their preferential susceptibility to oxidative stress. In Parkinson's disease, an overproduction of free radicals, suggested by the increased level of lipid peroxidation in the substantia nigra, might accelerate the rate of dopaminergic cells death. Impairment of the oxygen toxicity protective mechanisms may be responsible. Besides, the above normal levels of iron in the substantia nigra may contribute to free radical production and have a role in the toxic process, though one cannot exclude the increased iron content may be a non-specific product of cellular degeneration, a consequence of the gliosis.

Loss of midbrain dopaminergic neurons is a major pathological characteristic of Parkinson's disease (Agid et al., 1987). The different dopaminergic cell groups do not all degenerate to the same extent, the neuromelanin-pigmented neurons are preferentially affected (Hirsch et al., 1988), indicating dopaminergic neurons do not all have the same vulnerability to the disease. Present knowledge suggests neuronal death associated to

Parkinson's disease is related to a toxicity of endogenous or exogenous origin, possibly associating hyperoxidation phenomena. The hypothesis for a toxic factor has been supported by observation of cases of permanent parkinsonism diagnosed in man intoxicated with the selective dopaminergic neurotoxin 1-methyl-4-phenyl-1,2,3,6-tetrahydropyridine (MPTP) (Langston et al., 1983), and by search for environmental risk factors (Koller et al., 1990). Indeed a limited role for heredity is suggested by the low concordance rate in monozygotic twins (Duvoisin, 1986) though there might be a genetic component since autosomal dominant transmission in some families (Johnson et al., 1990) has been reported. The rate of clinical aggravation (Blin et al., 1990), of nigral cell loss (Mc Geer et al., 1977) and of striatal dopaminergic deficit (Schermann et al., 1989) suggests the pathological process progresses until decease of the patients, independently of normal aging-related neuronal loss. Thus, the cause of the insult is continuously present or has a slow ongoing consequence, disruption of a vital process. Indeed, the supranormal HLA-DR-positive microglia (McGeer et al., 1988) and basal level of lipid peroxidation (Dexter et al., 1989) in parkinsonian substantia nigra post mortem, are indicative of an enduring degenerative process.

The pattern of neuronal death in the mesencephalon of patients is consistent with a most severe neuronal loss always seen in the substantia nigra, and a relative sparing of other dopaminergic cell groups (Hassler, 1938; Forno, 1982; Jellinger, 1986; Hirsch et al., 1988; Gibb and Lees, 1991). Nevertheless, considerable individual variability is evident from patients to patients; if indeed, neuromelanin-pigmented neurons are more susceptible, melanized neurons remain visible in the substantia nigra at advanced stages or after long duration of the disease and non-pigmented neurons are also insulted. Differences among the characteristics of dopaminergic neuron populations may determine why some neurons degenerate and why others are preserved. The subpopulations of dopaminergic neurons may differ in phenotype conferring them a susceptibility to toxic agents, a capacity of facilitating the formation of constituents giving rise to toxic factors or a better protection against toxicity of endogenous or exogenous factors. Examination of the control or pathological substantia nigra points on factors in relation to hyperoxidation phenomena, and possibly candidates in modulating the rate of neuronal death related to Parkinson's disease. The origin of this putative overproduction of free radicals is not known, but three good candidates are neuromelanin, iron and dopamine itself.

Free radicals may be produced in large quantities in the human substantia nigra

Biochemical pathways leading to generation of superoxide, peroxide or hydroxyl radicals are especially active within midbrain dopaminergic neurons. Free radicals and quinones are produced in substantia nigra during the oxidative degradation of dopamine by monoamine oxidase. Free radicals are also formed during the synthesis of neuromelanin which is found in some dopaminergic neurons of the mesencephalon. The progressive accumulation of neuromelanin in these neurons may expose them to constant oxidative stress (Mann and Yates, 1983). Another source of free radicals in the substantia nigra is through iron which catalyses the formation of hydroxyl radicals from H_2O_2, the breakdown of lipid peroxides and accelerates non-enzymatic oxidation of a variety of molecules (Halliwell and Gutteridge, 1986).

Under normal conditions, the continuous production of free radicals is compensated for by powerful protective enzymes. Dismutation of super-oxide radical to hydrogen peroxide is catalyzed by superoxide dismutase (SOD). The resulting hydrogen peroxide generated by cellular metabolism is then decomposed to water by catalase or removed by glutathione peroxidase (GPX) which uses hydrogen peroxide to oxidize the reduced glutathione. Neuromelanin-pigmented neurons in the control human mesencephalon have specific characteristics with regards to this defence system, and the biochemical pathways leading to oxygen species formation may be particularly active in these cells. In the substantia nigra these neurons express high levels of mRNA for CuZn SOD (Ceballos et al., 1990; Zhang et al., 1993) as well as CuZnSOD protein (Zhang and Javoy-Agid, in preparation). Thus the neurons the most vulnerable to the pathological process in Parkinson's disease may need a high CuZnSOD content to facilitate removal of superoxide radicals. Alternatively, a high cellular CuZnSOD by promoting hydrogen peroxide production might contribute to the vulnerability of these neurons. The neuronal localization of CuZnSOD is all the more interesting considering the cellular localization of GPX. In human mesencephalon, the protein is evidenced exclusively in astrocytic glial cells (Damier et al., 1993). Therefore, the oxygen defence system of the dopaminergic nigral neurons may require a neuron glial cell coupling. In control brains, the density of GPX positive cells was high in the dopaminergic areas spared in Parkinson's disease such as central grey substance, relatively low in the substantia nigra pars compacta most affected in the disease and intermediate in the ventral tegmental area and the peri-and retrorubral region (Damier et al., 1993). Therefore, dopaminergic neurons surrounded by a low density of glial cells may be less protected against deleterious action of free radicals, which may explain their preferential susceptibility to oxidative stress and accelerate their degeneration. In Parkinson's disease, an increase in density of GPX positive glials cells is observed. This increase is related to the severity of the neuronal loss (Damier et al., 1993) and may be a protection of the still surviving neurons, against an excess of toxic species. Changes in the activities of enzymes and substrates involved in free radical formation have been evidenced on tissue homogenates of substantia nigra: an increase in SOD-like activity (Martilla et al., 1988; Saggu et al., 1989), a marked reduction in glutathione levels (Perry and Yong, 1986) and of GPX activity (Ambani et al., 1975) with no change in glutathione transferase activity (Perry and Yong, 1986) as well a reduction in catalase activity (Ambani et al., 1975). Therefore oxygen toxicity protective mechanisms may be impaired in the substantia nigra of patients, and oxygen toxicity events related may participate in neuronal degeneration associated to the disease.

Free radicals may be produced in excess in the substantia nigra of parkinsonian patients

a) There are several reasons for believing that *neuromelanin contributes to dopaminergic cell death in Parkinson's disease*. Several dopamine-containing cell groups in the human mesencephalon have been evidenced i.e. within the substantia nigra pars compacta, in the peri- and retrorubral tegmental cell group A8 and ventral tegmental area, in the paramedium tegmentum and in the central grey substance (Hirsch et al., 1988). The neurons of these cell groups differ according to some of their biochemical characteristics. For example, two types of dopaminergic neurons have been identified regarding that they contain or not neuromelanin. Within the control human midbrain the percentage of melanized neurons varies from very high (in substantia nigra pars compacta), to intermediate (in the ventral tegmental region and the peri- and retrorubral region), and very low (in the central grey substance) levels. In Parkinson's disease, among total population of catecholaminergic neurons in the midbrain, the subpopulations of pigmented neurons were the most vulnerable. Indeed, the higher percentage of melanized neurons in the dopaminergic cell group, the greater the loss of neurons in that cell group, as judged by comparisons with controls. Thus Parkinson's disease does not affect all dopaminergic neurons, but it affects preferentially the neuromelanized neurons. Among melanized neurons, those with a higher neuromelanin content are more susceptible to Parkinson's disease (Kastner et al., 1992). The gradual accumulation of neuromelanin in neurons over decades (Mann and Yates, 1993) may contribute to the pathological process through the production of free radicals (Graham, 1979) or by binding and release of toxic agents (Marsden, 1983). These findings suggest that the presence of neuromelanin in a dopaminergic neuron may be related directly or indirectly to the pathological process underlying cell death in this pathology. Mechanisms proposed as links between neuromelanin content and vulnerability of dopaminergic neurons include the generation of free radicals in the synthesis of neuromelanin (Marsden, 1983; Graham et al., 1978; Graham, 1979) and the binding or release by neuromelanin of toxic agents (Marsden, 1983). In addition, accumulation of neuromelanin and the production of free radicals or release of toxic agents could be gradual. However, per se, the presence of neuromelanin is not the unique factor determining susceptibility of dopaminergic neurons. Indeed, in Parkinson's disease, the pattern of neuronal loss within the mesencephalon is quite variable from one patient to another. In addition, melanized neurons are spared and non-melanized neurons degenerate (Hirsch et al., 1992). Furthermore, in progressive supranuclear palsy (PSP), a parkinsonian syndrome also associated with a loss of dopaminergic nigrostriatal neurons, the cell loss affects melanized as well as non-melanized dopaminergic neurons with equal severity (Hirsch et al., 1988).

b) *Dopamine*. Degradation of dopamine by monoamine oxidase is source of free radicals (Grahams et al., 1978); the increased rate of dopamine metabolism in the surviving dopaminergic neurons (Agid et al., 1987) might be responsible for an increased production of free radicals and thus, for an oxidative stress. If the increase in free radical production due to increased dopamine turnover is not buffered by the scavenging enzymes (SOD, catalase, glutathione peroxidase) the compensatory hyperactivity of the dopaminergic neurones may become self destructive.

Iron may contribute to free radical production

A role of physiological metals has been suspected in the pathophysiology of several neurodegenerative diseases: copper for Wilson's disease (Scheinberg and Sternlief, 1984), aluminium in Alzheimer's disease (Birchall and Cappel, 1988), manganese for some parkinsonian syndromes (Barbeau, 1984). In Parkinson's disease, iron could participate in the mechanism underlying increased lipid peroxidation, since iron may contribute to free radical production (Halliwell and Gutteridge, 1986), and is selectively bound to neuromelanin to produce Fe^{2+}/melanin complex suspected to induce an oxidative stress (Ben-Shachar et al., 1991). Iron is found in high concentrations in the substantia nigra, and other areas of basal ganglia. In tissue homogenates of parkinsonian brain, the total iron content is enhanced in substantia nigra, although not in many other brain regions (caudate nucleus, putamen, cerebral cortex) (Earle, 1968; Dexter et al., 1991) (Fig. 1). Iron increase concerns the ferric and not the ferrous form (Riederer et al., 1989). An excessive

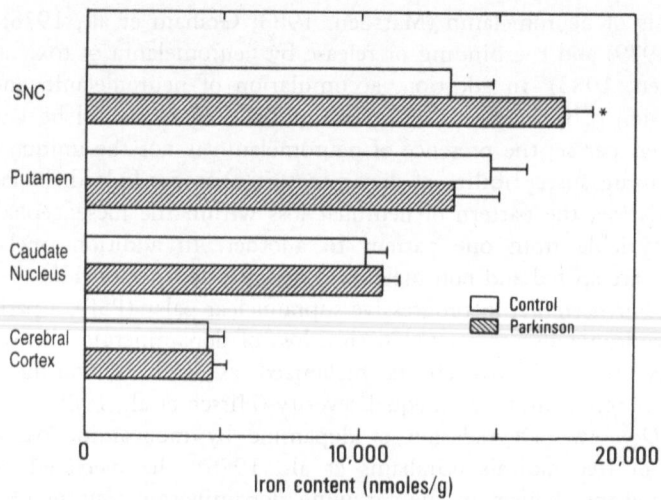

Fig. 1. Total levels of iron in parkinsonian and age-matched control brains. Values obtained on tissue homogenates of brain structures, represent mean ± SEM

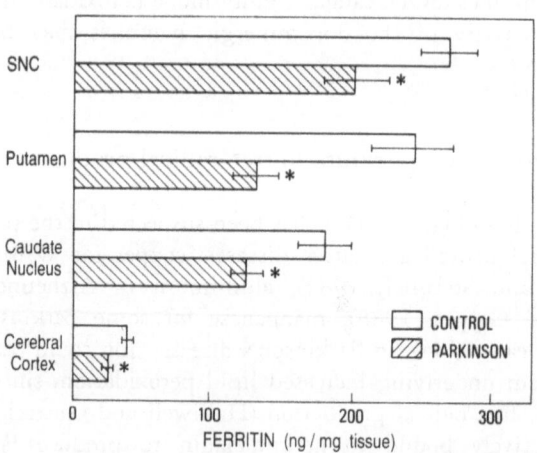

Fig. 2. Ferritin levels in parkinsonian and age-matched control brains. Data are the mean ± SEM of values obtained on tissue homogenates of brain structures

accumulation of iron in the nigra might accelerate cell death by aggravating the presence of free radicals at toxic levels. Not all forms of iron are capable of inducing free radical formation. Iron in a free form is toxic. In brain most iron is inactivated by association with ferritin (Hallgren and Sourander, 1958) with relatively little in a free and reactive form (Halliwell and Gutteridge, 1986). Ferritin in substantia nigra decreases in parkinsonian patients compared to controls suggesting a defective iron storage capacity (Dexter et al., 1991) (Fig. 2). Thus iron in excess may be in a free form. However, ferritin levels also decrease in other brain areas (Fig. 2) and other investigations reported above normal ferritin content in the nigra (Riederer et al., 1989). Further studies are required to understand whether the changes in brain iron and ferritin levels are related and are contributive factors to the cause of the disease. Why iron is increased is not known. The increase in nigral iron content may result from an increased transport of the metal into nigral cells as a transferrin-iron complex via specific cell surface receptors (Bomford and Munro, 1985). At present, there is no evidence for an alteration of iron transport capacity in the parkinsonian substantia nigra. Indeed, at macroscopic level the density of ferrotransferrin receptor binding sites was similar in control and pathological brains (Faucheux et al., 1993) (Fig. 3). The abnormal accumulation of iron in parkinsonian substantia nigra may be a non-specific product of cellular degeneration rather than a cause. The absence of increase in nigral iron content in patients with progressive

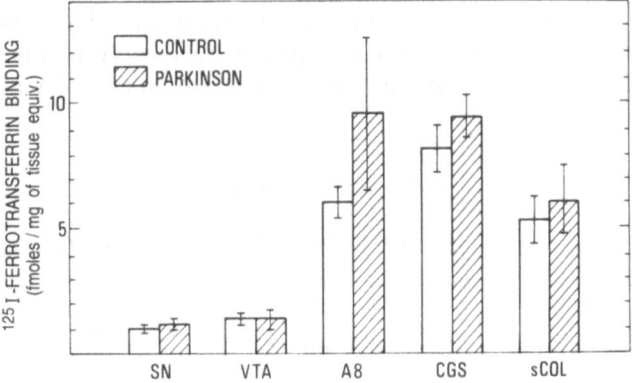

Fig. 3. Density of [125]-Iodo-ferrotransferrin receptors in the mesencephalon of control and parkinsonian brains as measured by autoradiographic detection on tissue sections. Data are the mean ± SEM of values obtained on 6 control and 4 parkinsonian brains. *SN* substantia nigra; *VTA* ventral tegmental area; *A8* peri- and retrorubral tegmental cell group; *CGS* central gray substance; *sCOL* superior colliculust

supranuclear palsy (Hirsch et al., 1991) suggests that the metal accumulation observed in parkinsonian substantia nigra is not merely a nonspecific marker of cell death. In addition, comparison of the nigral tissue content in controls and parkinsonians by X-ray microanalysis evidenced complex changes in iron levels: they increased in regions devoid of neuromelanin, but decreased in neuromelanin aggregates (Hirsch et al., 1991). Classically, iron is observed in astrocytes in the substantia nigra (Hallgren and Sourander, 1958), a gliosis could therefore explain the increased iron content in the substantia nigra of patients. The decreased iron content on neuromelanin aggregates may correspond to neurons that had low amounts of iron before the disease started (and explain their lesser vulnerability) or be a consequence of the pathological process on the remaining cells.

Finally, melanized dopaminergic neurons may be a site of preferential production of oxygen species, a result of polyfactorial mechanisms. Iron could be one of the factors. The increased iron content can catalyze oxidative stress, the metal would have a role in the toxic process. However, the abnormal accumulation of iron may be a consequence of cell degeneration.

Factors may in certain circumstances, potentiate there mutual effects, facilitating formation of species at toxic levels, thereby accelerating neuronal death to a pathological state. Their link with the cause of the disease (primary insult? one of the mechanisms participating a the final stage of cell death?) remains a mystery. The rate of cell death could also be modulated by the action of other neurotoxic or neuroprotective agents. Alone, these agents influence the susceptibility of the dopaminergic neurons, in combination, they may set the limits of vulnerability and accelerate the degradation process initiated by the pathogenic agent, leading ultimately to cell death, more rapidly than in normal aging.

Acknowledgements

This work was supported by INSERM and Association Claude Bernard.

References

Agid Y, Javoy-Agid F, Ruberg M (1987) Biochemistry of neurotransmitters in Parkinson's disease. In: Marsden CD, Fahn S (eds) Movement disorders, vol 2. Butterworth, London, pp 166–230

Ambani LM, Van Voert MH, Murphy S (1975) Brain peroxidase and catalase in Parkinson's disease. Arch Neurol 32: 114–118

Barbeau A (1984) Manganese and extrapyramidal disorders (a critical review and tribute to Dr. Georges C. Cotzias). Neurotoxicoloy 5: 13–36

Ben-Shachar D, Riederer P, Youdim MBH (1991) Iron-melanin interaction and lipid peroxidation: implications for Parkinson's disease. J Neurochem 57: 1609–1614

Birchall JD, Cappell JS (1988) Aluminium, chemical physiology and Alzheimer's disease. Lancet ii: 1008–1010

Blin J, Bonnet, AM, Vidailhet M, Brandabur M, Agid Y (1991) Does aging aggravate parkinsonian disability? J Neurol Neurosurg Psychiatry 54: 780–782

Bomford AB, Munro HN (1985) Transferrin and its receptors: their roles in cell function. Hepathology 5: 870–875

Ceballos I, Lafon M, Javoy-Agid F, Hirsch E, Nicole A, Sinet PM, Agid Y (1990) Superoxide dismutase and Parkinson's disease. Lancet i: 1035–1036

Damier P, Hirsch E, Javoy-Agid F, Zhang P, Agid Y (1993) Glutathione peroxidase, glial cells and Parkinson's disease. Neuroscience 52: 1–6

Dexter D, Carter C, Javoy-Agid F, Lees AJ, Jenner P, Marsden CD (1989) Basal lipid peroxidation in substantia nigra is increased in Parkinson's disease. J Neurochem 52: 381–389

Dexter DT, Carayon A, Javoy-Agid F, Agid Y Wells FR, Daniel SE, Lees AJ, Jenner P, Marsden CD (1991) Alterations in the levels of iron, ferritin and other trace metals in Parkinson's disease and other neurodegenerative diseases affecting the basal ganglia. Brain 114: 1953–1975

Duvoisin RC (1986) Etiology of Parkinson's disease: current concepts. Clin Neuropharmacol [Suppl 1]: S3–S11

Earle KM (1968) Studies on Parkinson's disease including X-ray fluorescent spectroscopy of formalin fixed brain tissue. J Neuropathol Exp Neurol 27: 1–14

Faucheux BA, Hirsch EC, Villares J, Selimi F, Mouatt-Prigent A, Javoy-Agid F, Hauw JJ, Agid Y (1993) Distribution of ferrotransferrin binding sites in the mesencephalon of control subjects and patients with Parkinson's disease. J Neurochem 60: 2338–2341

Forno LS (1982) Pathology of Parkinson's disease. In: Marsden CD, Fahn S (eds) Movement disorders. Butterworth, London, pp 25–40

Gibb WRG, Lees AJ (1991) Anatomy, pigmentation, ventral and dorsal subpopulations of the substantia nigra, and differential death in Parkinson's disease. J Neurol Neurosurg Psychiatry 54: 388–396

Graham DG (1979) On the origin and significance of neuromelanin. Arch Pathol Lab Med 103: 359–362

Graham DG, Tiffany SM, Bell WR (1978) Auto-oxidation versus covalent binding of quinones as the mechanism of toxicity of dopamine, 6-hydroxydopamine, and related compounds towards C 1300 neuroblastoma cells in vitro. Mol Pharmacol 14: 644–653

Hallgren B, Sourander P (1958) The effect of age on non haemin iron in human brain. J Neurochem 3: 41–51

Halliwell B, Gutteridge JMC (1986) Iron and free radical reactions: two aspects of antioxidant protection. Trends Biol Sci 11: 1372–1375

Hassler R (1938) Zur Pathologie der Paralysis agitans und des postencephalitischen Parkinsonismus. J Psychol Neurol 48: 387–476

Hirsch EC, Graybiel AM, Agid Y (1988) Melanized dopaminergic neurons are differentially affected in Parkinson's disease. Nature 334: 345–348

Hirsch EC, Brandel JP, Galleg P, Javoy-Agid F, Agid Y (1991) Iron and aluminium increase in the substantia nigra of patients with Parkinson's disease: an X-ray microanalysis. J Neurochem 56: 446–451

Hirsch EC, Mouatt A, Faucheux B, Bonnet AM, Javoy-Agid F, Graybiel A, Agid Y (1992) Dopamine, tremor and Parkinson's disease. Lancet i: 340: 125–126

Jellinger K (1986) Pathology of parkinsonism. In: Fahn S, Marsden CD, Jenner P, Teychenne P (eds) Recent development in Parkinson's disease. Raven Press, New York, pp 303–307

Johnson WG, Hodge SE, Duvoisill R (1990) Twin studies and the genetic of Parkinson's disease – a reappraisal. Mov Disord 53: 187–194

Kastner A, Hirsch EC, Lejeune O, Javoy-Agid F, Rascol O, Agid Y (1993) Is the vulnerability of neurons in the substantia nigra of patients with Parkinson's disease related to their neuromelanin content? J Neurochem 59: 1080–1092

Koller W, Vetere-Overfield B, Gray RN, Alexander BS, Chin T, Dolezal J, Hassanein R, Tanner C (1990) Environmental risk factors in Parkinson's disease. Neurology 40: 1218–1221

Langston JW, Ballard P, Tetrud JW, Irwin I (1983) Chronic parkinsonism in humans due to a product of mepedrine analogue synthesis. Science 219: 979–980

Mann DMA, Yates PO (1983) Possible role of neuromelanin in the pathogenesis of Parkinson's disease. Mech Ageing Dev 21: 193–203

Marsden CD (1983) Neuromelanin and Parkinson's disease. J Neural Transm [Suppl] 19: 121–141

Marttila RJ, Lorentz H, Rinne UK (1988) Oxygen toxicity protecting enzymes in Parkinson's disease. J Neurol Sci 86: 321–331

Mc Geer PL, Mc Geer EC, Suzuki J (1977) Aging and extrapyramidal function. Arch Neurol 34: 33–35

Mc Geer PL, Itagaki Q, Akiyama H, Mc Geer EG (1988) Rate of cell death in parkinsonism indicates active neuropathological process. Ann Neurol 24: 564–576

Perry TL, Yong WW (1986) Idiopathic Parkinson's disease, progressive supranuclear palsy and glutathione metabolism in the substantia nigra of patients. Neurosci Lett 67: 269–274

Riederer P, Sofic E, Rausch WD, Schmidt B, Reynolds GP, Jellinger K, Youdim MBH (1989) Transition metals, ferritin, glutathione and ascorbic acid in parkinsonian brains. J Neurochem 52: 515–520

Saggu H, Cooksey J, Dexter D, Wells FR, Lees A, Jenner P, Marsden CD (1989) A selective increase in particulate superoxide dismutase activity in parkinsonian substantia nigra. J Neurochem 53: 692–697

Scheinberg IH, Strenlief I (1984) Wilson's disease. Saunders, Philadelphia

Schermann D, Desnos C, Darchen F, Pollak P, Javoy-Agid F, Agid Y (1989) Striatal dopamine deficiency in Parkinson's disease: role of aging. Ann Neurol 26: 551–557

Youdim MBH, Ben-Shachar D (1989) Iron-melanin interaction in substantia nigra as the neurotoxic component of Parkinson's disease. II International Conference "Basic and Therapeutic Strategies XX Alzheimer's and Parkinson's disease" Kyoto, Japan (Abstracts, p 53)

Zhang P, Damler P, Hirsch EC, Agid Y, Ceballos-Picot I, Sinet PM, Nicole A, Laurent M, Javoy-Agid F (1993) Preferential expression of superoxide dismutase messenger RNA in melanized neurons in human mesencephalon. Neuroscience 55: 167–175

Correspondence: Dr. F. Javoy-Agid, Laboratoire de Medecine Experimentale, INSERM U 289, 47, boulevard de l'Hopital, F-75651 Paris Cedex 13, France.

Iron and neurotoxin intoxication: comparative in vitro and in vivo studies

W. Wesemann[1], St. Blaschke[1], H.-W. Clement[1], Chr. Grote[1], N. Weiner[1], W. Kolasiewicz[2], and K.-H. Sontag[2]

[1] Institute of Physiological Chemistry, Department of Neurochemistry, Philipps-University, Marburg, and [2] Max-Planck-Institute of Experimental Medicine, Göttingen, Federal Republic of Germany

Summary

The effect of iron on rat brain was studied in vivo in the nigrostriatal system and in vitro by assay of the lipid peroxidation in membranes isolated from cortex, hippocampus, and striatum. After intranigral injection of Fe (III) a reduced dopamine (DA) metabolism was observed in the striatum by in vivo pulse voltammetry. A retrograde lesion of the substantia nigra by intrastriatal injection of 6-hydroxydopamine (6-OHDA) could be enhanced by co-injection of iron. The lipid peroxidation as measured by ascorbate-induced malondialdehyde (MDA) production was increased by incubation with iron, whereas the iron chelator deferoxamine inhibited the formation of MDA.

Introduction

Though the etiology of Parkinson's disease is still unknown, many studies suggest the involvement of free radical-induced oxidative stress (Dexter et al., 1989; Riederer et al., 1989). The formation of free radicals from H_2O_2 can be increased by iron or other transition metals (Fenton reaction). These radicals can stimulate lipid peroxidation of cell membranes which can finally lead to cell death. The importance of Fe (III) for the generation of radicals and the increased Fe (III) levels found in the substantia nigra of parkinsonian patients may help to understand the selective degeneration of the dopaminergic nigrostriatal pathway (Dexter et al., 1989; Riederer et al., 1989; Hirsch et al., 1991; Sofic et al., 1991). Apparently the high iron levels of the substantia nigra are linked to the neuromelanin content of its pigmented cells. Neuromelanin forms iron complexes which were found to be increased in postmortem parkinsonian brains (Jellinger et al., 1992; Kastner et al., 1992).

In vitro as well as in vivo models are used to study the pathogenesis of parkinsonism and the effect of antiparkinsonian treatments. As was shown recently by HPLC, Fe (III) injection into the subtantia nigra of the rat reduces striatal catecholamine levels (Ben-Shachar and Youdim, 1991). In the present study we used pulse voltammetry as a sensitive in vivo method to analyse the striatal extracellular monoamine metabolite content (1) after unilateral Fe (III) injection into the substantia nigra and (2) after intrastriatal injection of a combination of 6-hydroxydopamine (6-OHDA) with Fe (III). To elucidate the possible mode of action of iron toxicity we studied in vitro the effect of iron and the iron chelator deferoxamine. Ascorbate-induced malondialdehyde (MDA) production was used to assay lipid peroxidation in three rat brain areas – cortex, hippocampus, and striatum.

Materials and methods

Voltammetry

Male Han-Wistar rats, weight 300–350 g, were anaesthetized with 400 mg chloralhydrate/kg, i.p., and used for dihydroxyphenylacetic acid (DOPAC) assay by in vivo voltammetry with carbon fiber electrodes (Ponchon et al., 1979; Gonon et al., 1980, 1981; Cespuglio et al., 1981). The electrodes were implanted in the striata of both hemispheres using the stereotactic coordinates of Paxinos and Watson (1982) – coordinates: A = + 8.5 mm; L = + 3.5 mm; V = + 3.5 mm. In acute experiments a solution of $FeCl_3$ in 2 μl NaCl (0.9%) was injected in the right substantia nigra pars compacta (coordinates: A = + 4.0 mm; L = + 1.9 mm; V = + 2.0 mm) using a glass capillary.

In chronic experiments the voltammetric analyses were performed one and three weeks after Fe (III) application. After 40 min registration 100 mg L-dihydroxyphenylalanine (DOPA)/kg i.p. were injected to test the restitution capacity of the nigrostriatum. In a second set of experiments 6-OHDA (8 μg/2 μl) was unilaterally injected with or without $FeCl_3$ (7.5 μg) into the ventrolateral striatum (coordinates: A = + 8.7 mm; L = + 4.0 mm; V = + 3.5 mm) to achieve a retrograde lesion of the substantia nigra. 3–5 months later the dopamine (DA) metabolism was determined by voltammetry.

Assay of lipid peroxidation in brain tissue homogenates

Crude membrane fractions were prepared from rat brain (cortex, hippocampus, striatum) as previously described (Weiner et al., 1992).

The assay of lipid peroxidation was based on methods published before (Heikkila et al., 1982; Villacara et al., 1989; Viani et al., 1991). Briefly, the samples (500 µl) containing 1000 µg protein/ml (Lowry et al., 1951) were incubated with 0.5 mM ascorbic acid at 37 °C for 30 min. The reaction was stopped with 20% trichloroacetic acid. After centrifugation (10,000 g, 15 min) 1 ml of thiobarbituric acid (0.67%) was added to 1 ml of the supernatant and the samples were incubated at 95 °C for 20 min. After centrifugation (10,000 g, 15 min), the amount of thiobarbituric acid reactive material (TBAR) as a measure of MDA production was spectrophotometrically quantified at 530 nm using 1,1,3,3-tetramethoxypropane (TMP) as standard (Buege and Aust, 1978). To test the effect of iron and the iron chelator deferoxamine on the ascorbate-induced MDA production we added to the incubation mixture with 0.5 mM ascorbic acid 0.01 mM $FeSO_4$ or 0.001/0.01 mM deferoxamine.

Results

Lesions with iron and 6-OHDA

When iron was injected unilaterally into the pars compacta of the substantia nigra, during the first three hours only rather high concentrations (200 µg/2 µl) showed a distinct reduction of extracellular DOPAC at the ipsilateral side when compared to the contralateral side (Fig. 1). Low concentrations (50 µg/2 µl) had no significant effect in acute experi-

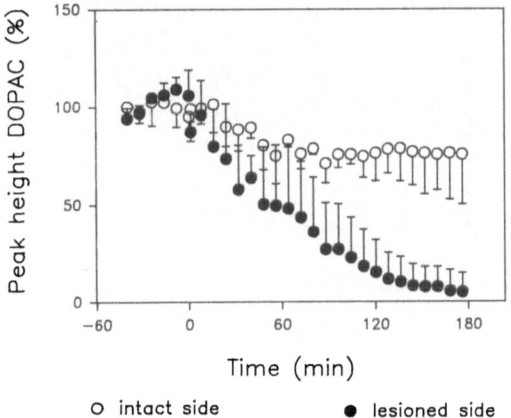

o intact side   ● lesioned side

Fig. 1. Acute effect of 200 µg $FeCl_3$ injected in the right substantia nigra pars compacta (at t = 0) of Wistar rats. The extracellular DOPAC concentration was measured by in vivo differential pulse voltammetry in the striatum. Concentrations are shown as % of the mean value from the 6 measurements before the injection (n = 4, ± S.E.M.)

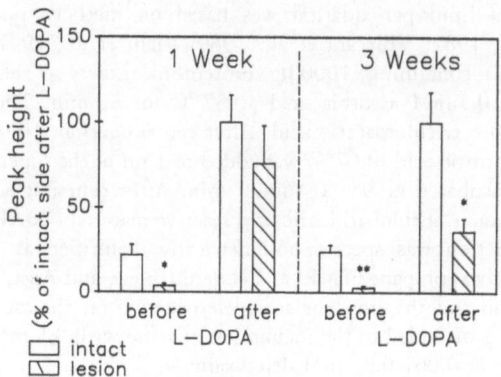

Fig. 2. Chronic lesion induced by injection of 50 µg FeCl₃ in the right substantia nigra pars compacta 1 or 3 weeks before assay of extraneuronal DOPAC in rat striatum by in vivo pulse voltammetry. L-DOPA was applied at a dose of 100 mg/kg. Concentrations are shown as % of intact side after L-DOPA (n = 4, ± S.E.M.)

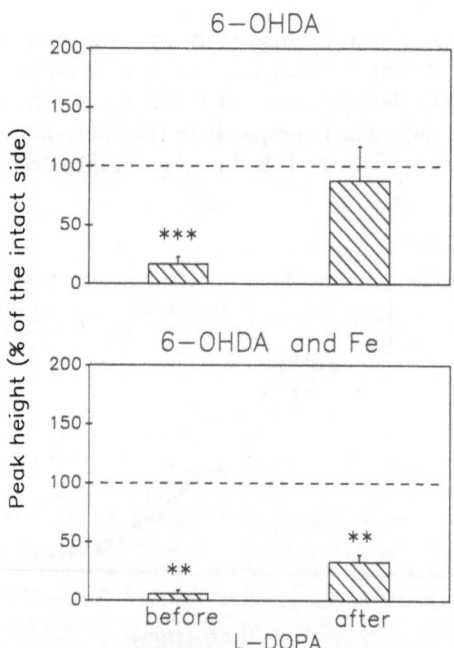

Fig. 3. Effect of a chronic lesion by striatal injection of 6-OHDA (8 µg/2 µl NaCl 0.9%) and 6-OHDA plus FeCl₃ (7.5 µg) on the extracellular DOPAC concentration in the striatum measured by in vivo pulse voltammetry 3–5 months after neurotoxin treatment. The restitution capacity was tested by application of L-DOPA (100 mg/kg .i.p). Concentrations at the lesioned side are shown as % of the intact side (n = 7–8, ± S.E.M.)

ments (result not shown). In chronic experiments, however, the same low iron concentration of 50 μg/2 μl induced a significant decrease in DOPAC peak height on the lesioned side as compared with the intact side 1 and 3 weeks after application (Fig. 2). After DOPA application the DOPAC peak was fully restored one week after the lesion but was significantly reduced after 3 weeks as compared with the control side. Figure 3 shows that the lesion observed 3–5 months after intrastriatal 6-OHDA (8 μg/2 μl) application can be intensified by co-administration of a rather low iron concentration (7.5 μg/2 μl). The difference between 6-OHDA and 6-OHDA + Fe (III) treated rats is even more distinct if the DOPA mediated restoration of the dopaminergic system were compared with each other. In the Fe (III) treated rats the DOPAC peak was only partly restored when compared with the control.

Lipid peroxidation

Incubation with 0.5 mM ascorbate produced different amounts of TBAR material in the 3 brain areas investigated (cortex < hippocampus < striatum). After addition of 0.01 mM $FeSo_4$ the MDA production was significantly higher and reached about the same level in all 3 areas (Fig. 4). 0.01 mM deferoxamine inhibited the ascorbate-induced MDA

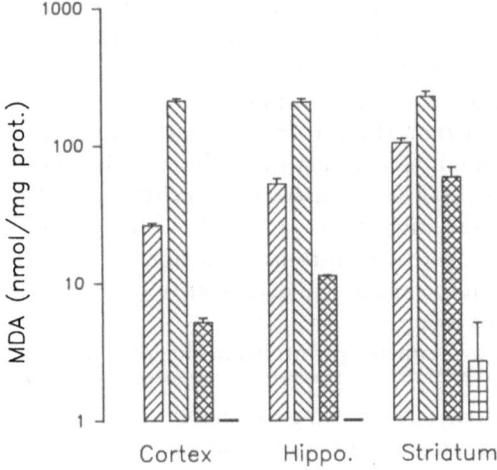

Fig. 4. Ascorbate-induced lipid peroxidation. Lipid peroxidation was assayed by measurement of malondialdehyde (MDA) formation in rat cortex, hippocampus and striatum homogenates. ▨ 0.5 mM ascorbate (n = 8–10); ▧ 0.5 mM ascorbate, 0.01 mM $FeSO_4$ (n = 5–7); ▨ 0.5 mM ascorbate; 0.001 mM deferoxamine (n = 4); ▨ 0.5 mM ascorbate; 0.01 mM deferoxamine (n = 4); results are means ± S.E.M

production almost completely in the 3 brain areas, whereas at 0.001 mM concentration the inhibition in the striatum was only 50%, in hippocampus and cortex about 80%.

Discussion

Differential pulse voltammetry and the assay of ascorbate-induced MDA production were used to study in vivo and in vitro the effects of iron on the nigrostriatal system of the rat and on lipid peroxidation. Apparently Fe (III)-induced neurotoxicity develops progressively: (1) Extremely high Fe (III) levels (200 $\mu g/2$ μl) were necessary to observe an acute lesion. (2) The amount of 50 μg Fe (III)/2 μl produced a lesion only after 1 week and even more pronounced after 3 weeks. These results are in accordance with data published by Ben Shachar and Youdim (1991) demonstrating the neurotoxic effect of iron on dopaminergic neurons of the rat. Though in agreement with the literature, the iron concentration of 50 $\mu g/2$ μl is still rather high. But one should keep in mind that as compared with primates rats are avoid of neuromelanin containing cells in the substantia nigra, and it is the neuromelanin which binds Fe (III) (Ben Shachar et al., 1991b). In addition, no H_2O_2 and/or radical producing neurotoxin was present in these experiments.

In order to provide a better model of parkinsonism in the rat a combination of iron with neurotoxin was applied. Iron was used to substitute the neuromelanin-iron complex, the neurotoxin 6-OHDA to generate radicals. In these experiments a retrograde degeneration of the nigrostriatal neurons was achieved by injection of the neurotoxin plus Fe (III) in the striatum (Berger et al., 1991; Cadet et al., 1991). In accordance with our assumption the addition of low iron concentrations (7.5 $\mu g/2$ μl) to 6-OHDA produced a further decrease of the DOPAC signal if compared with rats treated with 6-OHDA only. These in vivo results could mean that in parkinsonism iron accumulation by itself or low iron levels in the presence of degenerative radical producing processes play a hazardous role.

The in vitro experiments performed to elucidate the mechanism of iron-induced neurotoxicity showed that iron increased the lipid peroxidation of ascorbate drastically. This finding supplies further evidence indicating the role of oxidative stress for the etiology of parkinsonism and other diseases (Dexter et al., 1989; Riederer et al., 1989; Youdim et al., 1990; Halliwell, 1992). The incubation with the iron chelator deferoxamine inhibited almost completely the ascorbat-induced MDA production in the presence and absence of iron. It has been demonstrated that deferoxamine is also able to protect DA neurons from 6-OHDA-induced degeneration (Ben

Shachar et al., 1991a). The inhibiting effect on a radical-inducing neurotoxin and on lipid peroxidation supports the hypothesis that iron chelators may be useful in the prevention of parkinsonism.

References

Ben-Shachar D, Youdim MBH (1991) Intranigral iron injection induces behavioral and biochemical "Parkinsonism" in rats. J Neurochem 57: 2133–2135

Ben-Shachar D, Eshel G, Finberg JPM, Youdim MBH (1991a) The iron chelator desferrioxamine (desferal) retards 6-hydroxydopamine-induced degeneration of nigrostriatal neurons. J Neurochem 56: 1441–1444

Ben-Shachar D, Riederer P, Youdim MBH (1991b) Iron melanin interaction and lipid peroxidation: implications for Parkinson's disease. J Neurochem 57: 1609–1614

Berger K, Przedborski S, Cadet JL (1991) Retrograde degeneration of nigrostriatal neurons induced by intrastriatal 6-hydroxydopamine injection in rats. Brain Res Bull 26: 301–307

Buege JA, Aust SD (1978) Microsomal lipid peroxidation. In: Fleisher S, Packers L (eds) Methods in enzymology, vol 52. Biomembranes. Academic Press, New York, p 90

Cadet JL, Reuben L, Kostic V, Przedborski S, Jackson-Lewis V (1991) Long-term behavioral and biochemical effects of 6-hydroxydopamine injections in rat caudate-putamen. Brain Res Bull 26: 707–713

Cespuglio R, Faradji H, Ponchon JL, Buda M, Riou F, Gonon F, Pujol JF, Jouvet M (1981) Differential pulse voltammetry in brain tissue. I. Detection of 5-hydroxyindoles in the rat striatum. Brain Res 223: 287–298

Dexter DT, Wells FR, Lees AJ, Agid F, Agid Y, Jenner P, Marsden CD (1989) Increased nigral iron content and alterations in other metal ions occurring in brain in Parkinson's disease. J Neurochem 52: 1830–1836

Gonon F, Buda M, Cespuglio R, Jouvet M, Pujol JF (1980) In vivo electrochemical detection of catechols in the neostriatum of anaesthetized rats: dopamine or DOPAC? Nature 286: 902–904

Gonon FG, Fombarlet CM, Buda MJ, Pujol JF (1981) Electrochemical treatment of pyrolytic carbon fibre electrodes. Anal Chem 53: 1386-1389

Halliwell B (1992) Reactive oxygen species and the central nervous system. J Neurochem 59: 1609–1623

Heikkila RE, Cabbat FS, Manzino L (1982) Inhibitory effects of ascorbic acid on the binding of [^{3}H]dopamine agonists to neostriatal membrane preparations: relationship to peroxidation. J Neurochem 38: 1000–1006

Hirsch EC, Brandel JP, Galle P, Javoy-Agid F, Agid Y (1991) Iron and aluminium increase in the substantia nigra of patients with Parkinson's disease: an X-ray microanalysis. J Neurochem 56: 446–451

Jellinger K, Kienzl E, Rumpelmair G, Riederer P, Stachelberger H, Ben-Shachar D, Youdim MBH (1992) Iron-melanin complex in substantia nigra of parkinsonian brains: an X-ray microanalysis. J Neurochem 59: 1168–1171

Kastner A, Hirsch EC, Lejeune O, Javoy-Agid F, Rascol O, Agid Y (1992) Is the vulnerability of neurons in the substantia nigra of patients with Parkinson's disease related to their neuromelanin content? J Neurochem 59: 1080–1089

Lowry OH, Rosebrough NJ, Farr AL, Randall RJ (1951) Protein measurement with the Folin phenol reagent. J Biol Chem 193: 265–275

Paxinos G, Watson C (1982) The rat brain in stereotaxic coordinates. Academic Press, New York

Ponchon JL, Cespuglio R, Gonon F, Jouvet M, Pujol JF (1979) Normal pulse polarography with carbon fiber electrodes for in vitro and in vivo determination of catecholamines. Anal Chem 51: 1483–1486

Riederer P, Sofic E, Rausch WD, Schmidt B, Reynolds GP, Jellinger K, Youdim MBH (1989) Transition metals, ferritin, glutathione, and ascorbic acid in Parkinsonian brains. J Neurochem 52: 515–520

Sofic E, Paulus W, Jellinger K, Riederer P, Youdim MBH (1991) Selective increase of iron in substantia nigra zona compacta of Parkinsonian brains. J Neurochem 56: 978–982

Viani P, Cerrato G, Fiorilli A, Cestaro B (1991) Age-related differences in synaptosomal peroxidative damage and membrane properties. J Neurochem 56: 253–258

Villacara A, Kumami K, Yamamoto T, Mrsulja BB, Spatz M (1989) Ischemic modification of cerebrocortical membranes: 5-hydroxytryptamine receptors, fluidity and inducible in vitro lipid peroxidation. J Neurochem 53: 595–601

Weiner N, Clement H-W, Gemsa D, Wesemann W (1992) Circadian and seasonal rhythms of 5-HT receptor subtypes, membrane anisotropy and 5-HT release in hippocampus and cortex of the rat. Neurochem Int 21: 7–14

Youdim MBH, Ben-Shachar D, Yehuda S, Riederer P (1990) The role of iron in the basal ganglion. Adv Neurol 53: 155–162

Correspondence: Prof. Dr. W. Wesemann, Institut für Physiologische Chemie, Philipps-Universität, Hans-Meerwein-Strasse, D-35033 Marburg, Federal Republic of Germany.

Intranigral iron infusion in rats: A progressive model for excess nigral iron levels in Parkinson's disease?

G. W. Arendash[1], C. W. Olanow[2], and G. J. Sengstock[1]

[1] Department of Biology and Institute for Biomolecular Science, and
[2] Departments of Neurology, Pharmacology, and Psychiatry, University of South Florida, Tampa, Florida, U.S.A.

Summary

Increased iron levels are present within the substantia nigra zona compacta (SNc) of Parkinson's-diseased (PD) brains. Neurodegeneration within the SNc may be a direct consequence of increased iron because of iron's ability to catalyze cytotoxic free radical formation. We have developed an animal model based on excess nigral iron in PD through intranigral infusion of iron citrate in rats. These infusions cause a long-term and dose-related increase in nigral iron, with accompanying dose-related neuronal loss/gliosis within SNc and associated decreases in striatal dopaminergic markers. Moreover, intranigral iron infusion provides several long-term "progressive" changes including: 1) a progressive decrease in striatal dopamine and HVA, 2) a progressive atrophy of the substantia nigra, and 3) a progressive increase in apomorphine-induced rotational behavior. Since nigral lipid peroxidation is elevated following nigral iron infusion, free radical formation and ensuing oxidative damage may be involved in iron's neurotoxicity within the substantia nigra. This intranigral iron infusion model could be of extraordinary value in testing antioxidants for treating/preventing PD.

Introduction

The severe motor dysfunctions of Parkinson's disease (PD) occur primarily because of an accentuated, progressive loss of nigrostriatal dopaminergic (NS-DA) neurons, resulting in progressive decreases in striatal dopamine to levels insufficient for normal motor function. Concern that NS-DA neurons may be particularly vulnerable to "oxidant stress" has been raised due to the manner in which dopamine is metabolized (Olanow, 1992). Dopamine is oxidized to form a pool of hydrogen peroxide, either through enzymatic reaction with monoamine oxidase (MAO) or through nonenzymatic "auto-oxidation" (Fig. 1): Although this hydrogen peroxide is normally detoxified through oxidation of glutathione by glutathione peroxidase, the intraneuronal glutathione pool is saturable. Excess or increased steady state levels of hydrogen peroxide could react with ferrous iron to generate highly reactive hydroxyl free radicals (Fig. 1). Hydroxyl free radicals appear to be the primary mediators of oxidative damage through their initiation/induction of cytotoxic processes such as lipid peroxidation and DNA damage (Halliwell and Gutteridge, 1989). An increase in lipid peroxidation has, in fact, been reported within the SN of postmortem PD brains (Dexter et al., 1989a). Thus, oxidative metabolism of DA has the potential to generate toxic free radicals that could be directly responsible for the death of NS-DA neurons. Oxidant-induced neurodegeneration may be particularly relevent in the Parkinsonian brain because:

1. DA metabolism (turnover) is increased in surviving NS-DA neurons of PD brains (Zigmond et al., 1990), which presumable results in greater intraneuronal production of hydrogen peroxide.

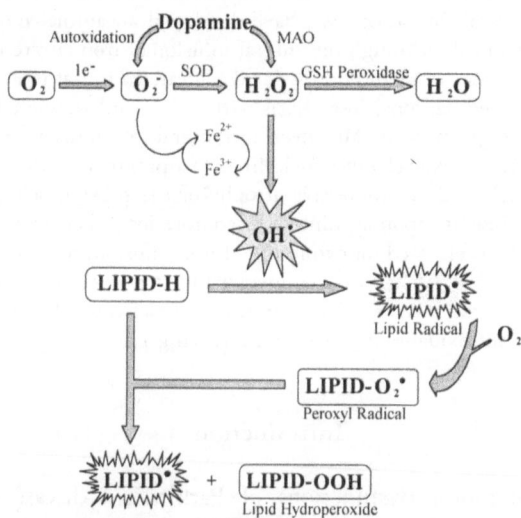

Fig. 1. Diagram indicating how the metabolism of dopamine, through autoxidation or monoamine oxidase (MAO), can result in increased hydroxyl free radical formation and consequent lipid peroxidative damage to cell membranes. Note the key role of iron in catalyzing conversion of excess hydrogen peroxide to hydroxyl free radicals. *GSH* reduced glutathione; *SOD* supraoxide dismutase

2. Reduced levels of glutathione have been reported in the SN of PD brains (Perry and Yong, 1986), which suggests decreased protection against nigral oxidant damage.
3. The SN of PD brains has an increased concentration of iron, which may catalyze increased formation of cytotoxic hydroxyl free radicals.

Separately, or in combination, the above characteristics of the Parkinsonian brain could provide a chronic "oxidative stress", resulting in the exacerbated progressive loss of NS-DA neurons in PD. Of particular significance may be the consistent reports of increased iron levels within the SN of PD brains (Sofic et al., 1988; Dexter et al., 1989b; Riederer et al., 1989). Recent studies indicate that this increase in nigral iron levels occurs primarily within zona compacta (SNc; Sofic et al., 1991), not only in association with reactive astrocytes and microglia, but also in association with neuromelanin granules of pigmented SNc neurons (Jellinger et al., 1990, 1992; Good et al., 1992). Neuromelanin granules may confer vulnerability to SNc neurons because of their high affinity for, and accumulation of, iron (Youdim et al., 1989). Moreover, because iron bound to ferritin is in a relatively unreactive state for catalyzing free radical formation, it is noteworthy that ferritin levels in the SN of PD brains have been reported to be significantly decreased (Dexter et al., 1990), suggesting the presence of excess "free or low-molecular weight" iron – iron that has the potential to catalyze formation of cytotoxic free radicals. By providing an oxidant stress, excess nigral iron may function as an endogenous neurotoxin and, thus, a primary etiological factor in the pathogenesis of PD (Youdim et al., 1989).

In view of the distinct possibility that excess nigral iron is toxic to NS-DA neurons in PD, we have sought to experimentally mimic this abnormal accumulation of nigral iron through the infusion of a recently-developed iron citrate solution (Sengstock et al., 1992) into the rat SN. This "intranigral iron infusion" model for PD has been shown to provide several aspects of PD not apparent in either the 6-OH DA or MPTP models including: 1. an increase in nigral lipid peroxidation, 2. progressive behavioral, neuropathological, and neurochemical changes that occur over months, and 3. a pathogenic mechanism that may more closely model the pathogenesis of PD.

Materials and methods

Iron citrate solutions prepared for intranigral infusion were all pH balanced and isosmotic to brain tissue (Sengstock et al., 1992, 1993a), thus assuring that any effects observed following infusion would be directly attributable to iron.

Adult male Sprague-Dawley rats received infusion(s) of either citrate vehicle
solution or iron citrate (Fe) solution at the junction between the substantia nigra's
zona compacta (SNc) and zona reticularis (SNr) according to one of the following
protocols: *single infusion*: a single 0.25 ul infusion of 0.63 nmol Fe^{+++} [35 ng Fe]
or vehicle solution was infused into the SN unilaterally; *side-by-side infusion*: two
side-by-side infusions (0.25 μl each) of vehicle or 0.63–3.15 nmol Fe^{+++} [70 ng–
350 ng total Fe] were infused unilaterally into the SN during the same surgical
session (Sengstock et al., 1992, 1993a).

Single infusion animals were sacrificed between one hour and one month
postinfusion. For histologic analysis of the nigral infusion site, the SN region
from some of these animals was sectioned and alternate sections stained for
thionin or iron (Perl's stain + DAB intensification). For the remaining single
infusion animals, the substantia nigra was discretely dissected out bilaterally and
processed for analysis of malondialdehyde (MDA) levels, an intermediate of lipid
peroxidation. To prevent iron-catalyzed or nonspecific MDA formation following
decapitation, both desferrioxamine and butylated hydroxytoluene (BHT) were
incorporated into our spectrophotometric MDA microassay. This methodology
results in a measurement of *basal* MDA levels because all available iron is
chelated.

Side-by-side infusion animals were tested monthly for apomorphine-induced
rotational behavior (0.25 mg/kg, s.c.) and were sacrificed one to six months
postinfusion. The SN region from each animal was processed histologically and
alternate brain sections stained for thionin or iron. Bilateral neostriatal tissues
were dissected out and assayed for biogenic amines and their catabolites by high
performance liquid chromatography (HPLC) with electrochemical detection.

Results

Single infusion studies

At one day following single infusion of 0.63 nmol Fe into the SN
unilaterally, the SNc contralateral to Fe infusion exhibited essentially no
iron staining, as is typical for this nigral region (Fig. 2, control uninfused
side). However, the Fe-infused SNc showed substantial iron staining;
although some of this enhanced staining was associated with glial cells,
much of it was apparently localized to neurons. Characteristic of these
iron-positive SNc neurons (or neuron-like cells) was a lightly-stained
cytoplasm and an intensely stained nucleus containing a variable number
of large iron granules and a very heavily stained nucleolus. By one month
postinfusion, iron-positive SNc neurons were much less evident and,
perhaps relatedly, substantial SNc neuronal loss/gliosis was present within the Fe-infused SNc region. Thus, infused Fe may associate directly
with SNc neurons to directly induce their degeneration.

At one hour following single infusion of 0.63 nmol Fe into the SN,

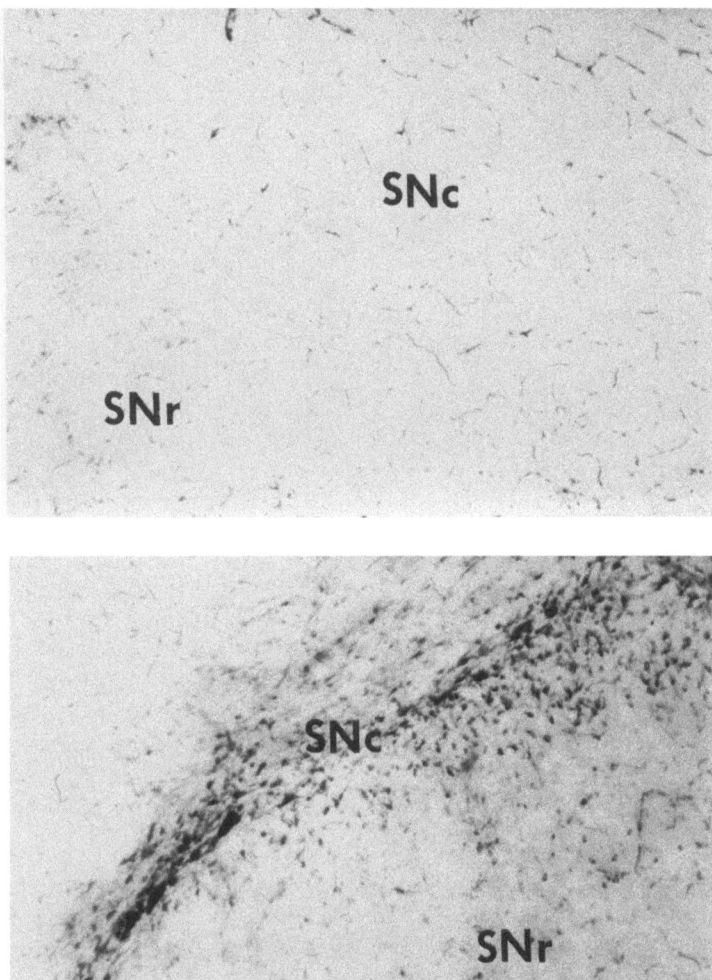

Fig. 2. Photomicrographs of an iron-stained coronal section through the substantia nigra at one month following side-by-side infusion of 1.25 nmol Fe (LOW Fe) at the interface between zona compacta (*SNc*) and zona reticularis (*SNr*). Compared to the minimal SNr staining and lack of SNc staining on the uninfused side (**upper**), intense iron staining is evident within the iron-infused substantia nigra (**lower**); this is particularly true in the SNc, where most iron positive cells appear to be reactive microglia and reactive astrocytes. Magnification: ×100

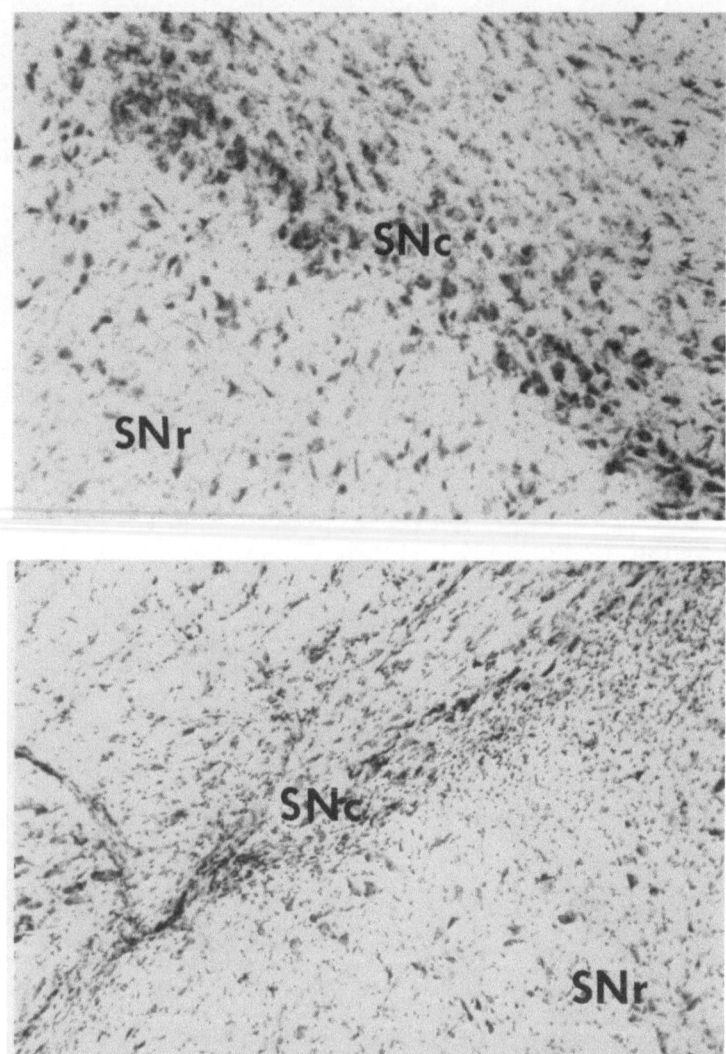

Fig. 3. An immediately adjacent thionine-stained brain section from the same animal in Fig. 2 that received side-by-side intranigral infusion of 1.25 nmol Fe one month prior to sacrifice. Extensive neuronal loss and gliosis are apparent within the iron-infused SNc (**lower**) compared to the contralateral uninfused SNc (**upper**) Note that the area of neuronal degeneration and gliosis corresponds with the area of increased iron staining (compare lower photomicrographs of Fig. 2 and 3). Magnification: ×100

MDA levels in the iron-infused SN were significantly elevated (by 72%) in comparison to values from unoperated control animals. Although MDA levels in other Fe-infused animals were elevated by 25% at 24 hours post-infusion, this elevation was not statistically significant and Fe-infusions did not affect nigral MDA levels in animals sacrificed at one week or one month after infusion. These neurochemical data indicate that Fe infusions into the SN acutely increase nigral lipid peroxidation and suggest that Fe-induced neurodegeneration within the SNc may be due, at least in part, to increased lipid peroxidation.

Side-by-side infusion studies

At 1–2 months following side-by-side infusion of 1.25 nmol total Fe (LOW Fe), a marked increase in iron staining intensity was evident and limited to the infused SN (Sengstock et al., 1992, 1993a). Heaviest iron staining was in the SNc and extreme dorsal SNr (Fig. 2). The Fe-infused SNc was clearly demarcated by large numbers of presumed reactive astrocytes and microglia, which stained intensely for iron (Fig. 2). Occassionally, iron-positive nigral neurons were observed (primarily in SNr), with a few SNc iron-positive neurons evident along the periphery of the zone staining intensely for iron. Corresponding closely to this area of intense iron staining was a reactive gliosis and extensive neuronal loss; this was particularly evident from analysis of adjacent thionin-stained brain sections (Fig. 3). It should be noted that nigral infusions of LOW Fe destroy less than half the population of SNc neurons. However, side-by-side infusion of increasing amounts of iron (2.10–6.3 nmol total Fe) resulted in an increasingly larger area of iron staining and increasingly more severe neuronal loss in the SN at two months (Sengstock et al., 1993a). Infusion of 2.10 nmol total Fe (MEDIUM Fe) destroyed the majority of SNc neurons and was the highest amount of Fe infused where Fe diffusion could be limited to the SN. Although infusion of higher Fe amounts (3.15–6.3 nmol; HIGH Fe) induced near complete destruction of SNc neurons, extranigral neurodegeneration was also evident at these doses.

Consistent with the dose-dependent destruction of SNc neurons 1–2 months following side-by-side Fe infusions were accompanying dose-related decreases in ipsilateral striatal concentrations of DA, DOPAC, and HVA compared to the contralateral uninfused side (Sengstock et al., 1992, 1993a); highly significant linear correlation values of −0.88, −0.79, and −0.87, respectively, resulted with infusion of increasing iron amounts. For example, striatal DA levels were significantly decreased by 24–44% at two months following MEDIUM Fe infusions and by 82%

after HIGH Fe infusions. It is noteworthy that the decreases in striatal DA and its catabolites induced by LOW Fe infusions were not significant at 1–2 months postinfusion; this is most probably because less than half the SNc neuronal population was destroyed. Nondopaminergic markers in striatum (i.e., norepinephrine, 5-HT, 5-HIAA) were essentially unaffected by any amount of iron infused, thus indicating a degree of neurochemical specificity and affirming that iron-induced neurodegeneration was restricted to the SN region.

Progressive changes following side-by-side iron infusions

All of the aforementioned histological and neurochemical effects of intranigral iron infusion were attained from animals sacrificed two months or earlier after iron infusion. To elucidate any iron-induced *progressive* changes occurring slowly over a more protracted period of time, it is necessary to determine the time-course of neuropathologic, neurochemical, and behavioral effects resulting from nigral iron infusion over many months. Of fundamental importance in such a progression study is the selection of an iron dose that creates only moderate initial loss of NS-DA neurons in conjunction with modest depletions in striatal DA markers. Side-by-side infusion of 1.25 nmol Fe (LOW Fe) or 2.10 nmol Fe (MEDIUM Fe) fulfill this requirement in terms of initial neuronal loss and depletion of striatal DA markers (less than 50% at 2 months postinfusion). Therefore, both of these doses were utilized in a time course study involving sacrifice at 2, 4, or 6 months following intranigral iron infusion (Sengstock et al., 1993b).

Irrespective of whether LOW or MEDIUM Fe was infused, a semi-quantitative analysis of SNc neuronal numbers in thionin-stained brain sections revealed no greater neuronal loss at four or six months postinfusion compared to the moderate loss observed at two months. Consistent with this lack of progressive SNc neuronal loss after iron infusion, tyrosine hydroxylase (TH) immunohistochemical staining following infusion of either LOW or MEDIUM Fe revealed persistent decreases in TH immunoreactivity within the infused SN through six months. This persistent iron-induced decrease in TH staining was due to a decreased number of immunoreactive cells and fibers within the infused SNc. The relative selectivity of LOW and MEDIUM iron infusions for inducing neurodegeneration of NS-DA neurons is underscored by the lack of any effect on TH-positive neurons in the ventral tegmental area (an immediately adjacent population of DA neurons with mesolimbic projections). Although a progressive iron-induced loss of NS-DA neurons was not observed in this study, a computer-assisted determination of SN volume from both LOW and MEDIUM Fe-infused animals indicated that a significant atrophy of the iron-infused SN had in fact occurred after 2 months. In the LOW Fe animals, the atrophy progressed through 4 months and in the MEDIUM Fe animals significant progression beyond the 2 month point was seen at 6 months postinfusion.

Intranigral iron infusions also have the potential to induce progressive decreases in neostriatal DA and HVA levels over a period of at least four months, although this progressive effect is in relation to values on the contralateral side and appears to be dose-dependent. Thus, a progressive decrease in striatal DA levels was *not* seen following MEDIUM Fe infusions because the significant 24% reduction in ipsilateral DA at two months postinfusion was not significantly intensified at the 4 or 6 month time points. By contrast, LOW Fe infusions did provide a progressive decrease in both striatal DA and HVA levels between 2 and 4 months postinfusion. For DA, this progressive decrease occurred because LOW Fe-infused animals sacrificed at two months postinfusion had a non-significant 8% decrease in striatal DA, while those sacrificed at the 4 and 6 month time points had significant 22% reductions in striatal DA levels. These neurochemical reductions are significant in several respects. Firstly, they represent the first truly progressive decreases in striatal DA and HVA levels achieved in an animal model for PD. Secondly, despite only modest iron-induced reductions in striatal DA and HVA levels at two months following either LOW or MEDIUM Fe infusion, no recovery occurred in either striatal DA or HVA levels at later time points.

In rodents, the behavioral endpoint most often used to access the completeness of lesions involving the NS-DA pathway is apomorphine-induced rotational behavior. Following extensive 6-OH DA induced lesions of the NS-DA system (i.e. unilateral lesions resulting in at least a 90% depletion in striatal DA levels ipsilaterally), apomorphine adminis-tration induces a significant degree of contralateral rotation – usually at least 7 or 8 complete rotations per minute. This effect is normally present within weeks of 6-OH DA lesioning and is not progressive thereafter, due in large part to the near complete initial loss of NS-DA neurons ipsilateral to 6-OH DA administration. Because iron-induced depletions in striatal DA and HVA levels never exceeded 32% of contralateral values throughout the present six month progression study, it was not surpris-ing to find no contralateral rotation in iron-infused animals during monthly test sessions for apomorphine-induced rotation behavior. How-ever, substantial *ipsilateral* rotation was observed in LOW and MEDIUM Fe-infused animals during these sessions. Moreoever, this ipsilateral rotation increased progressively throughout the six months of behavioral testing, irrespective of which iron amount had been infused (p < .005). The mechanism(s) responsible for the prominent and progressive ipsilateral rotation observed in LOW and MEDIUM Fe-infused animals following apomorphine will require further elucidation to determine if this is primarily an effect of SNc neuronal loss or an effect induced by loss of non-DA neurons within the SN.

Vehicle-infused animals

To control for any mechanical/volumetric effects resulting from intranigral iron infusion, all studies involved infusion of citrate vehicle solution alone (pH 7 and isosmotic). These infusions had no significant effect on any neuropathologic, neurochemical, or behavioral endpoint that was evaluated, indicating that effects seen following nigral iron infusion were directly attributable to iron.

Discussion

Iron has been suggested to contribute to the pathogenesis of PD because iron levels are increased in the substantia nigra of PD brains, specifically within the zona compacta (SNc; Sofic et al., 1991). The progressive degeneration of NS-DA neurons within the Parkinsonian SNc may be, in part, a consequence of excess "free or low molecular weight" iron, which promotes cytotoxic free radical formation (Olanow, 1992). Thus, PD may involve a disruption of brain iron homeostasis. Over the past 3–4 years, we have begun to elucidate the neuropathologic, neurochemical, and behavioral changes induced by excess iron within the rat SN and, in so doing, have begun to define a new "intranigral iron infusion model" for PD which could reflect the disease's etiology and/or pathogenesis more closely than other animal models currently being utilized.

Acutely (i.e., 24 hours) following low-dose intranigral iron infusion, SNc neurons appear to stain intensely for iron, suggesting an intraneuronal uptake of infused iron. More sophisticated techniques such as laser microprobe mass analysis (LAMMA) would help to definitely establish this finding. Neuronal iron uptake could occur in association with transferrin via transferrin receptors located on cell membranes (Irie and Tavassoli, 1987) or through some as yet undefined transferrin-independent mechanism (Basset et al., 1986). By several weeks following low-dose iron infusion into the SN, the area of intense iron staining is localized to the SNc and dorsal-most portion of the SNr stain, wherein the vast majority of iron-positive cells appear to be reactive astrocytes and microglia (Sengstock et al., 1992, 1993a). Corresponding closely to this SN region of intense iron staining is an area of extensive neuronal loss and reactive gliosis, the latter of which persists at least through 6 months postinfusion (Sengstock et al., 1993b).

The above chronology of iron-induced pathological events suggests that infused iron, initially taken up by SNc neurons, induces neurodegeneration through iron-induced free radical formation/oxidative damage. An ensuing infiltration of reactive glia into the SNc would then appear to phagocytize iron-rich neuronal debris and to remain in the SNc for many months. It is noteworthy that this long-term distribution of cellular iron staining within the SNc of iron-infused rats (i.e., iron associated primarily with reactive glia) has also been reported to characterize the SNc of PD brains (Jellinger et al., 1990), suggesting that the SN of PD brains and of iron-infused rats share a similar pathological process involving iron dysregulation/excess. We have recently determined, however, that 6-OH DA lesions of the NS-DA pathway also result in increased SNc iron staining, as well as increased neurochemical iron concentrations within the SNc (Oestreicher et al., submitted). Although these findings suggest that increased iron in the SN of PD brains is a consequence of NS-DA neurodegeneration, increased nigral iron may still play an important role in the pathogenesis of PD – especially since iron released from degenerative NS-DA neurons is apparently taken up by neighboring NS-DA neurons (Oestreicher et al., submitted). Moreover, the intranigral iron infusion model appears capable of inducing increased lipid peroxidation within the SN – an important characteristic of PD (Dexter et al., 1989a). This finding is consistent with iron-induced free radical formation/ oxidative damage within the infused SN, although further studies are clearly necessary to establish an association between iron infusion and increased free radical formation/oxidative damage. Although the neurotoxic mechanisms of both 6-OH DA and MPTP have been hypothesized to involve oxygen free radical formation, to date, no reports of increased lipid peroxidation within the rodent SN following 6-OH DA infusion have appeared. Furthermore, demonstration of lipid peroxidation in vivo following MPTP administration has been unsuccessful (Corongiu et al., 1987; Kopin and Schoenberg, 1988).

With infusion of higher iron amounts (2.1–6.3 nmol Fe) intranigrally, a dose-related increase in nigral iron, as well as in extent of SNc neuronal loss and reactive gliosis, were evident (Sengstock et al., 1993a). Relatedly, a highly significant correlation exists between the amount of iron infused intranigrally and the magnitude of ipsilateral reductions in striatal DA, DOPAC, and HVA (Sengstock et al., 1993). We have also found that a significant correlation exists between the degree of iron-induced neuronal loss within SNc and the magnitude of ipsilateral decreases in striatal DA concentrations. Thus, the extent of iron-induced nigral damage correlates closely with the degree of reduction in striatal DA levels.

An ideal animal model for PD should not only reflect the disease's pathogenesis, but also provide the *progressive* loss of NS-DA neurons and associated *progressive* depletions in striatal DA that characterize PD. Although no animal model for PD has been demonstrated to induce progressive loss of NS-DA neurons, the intranigral iron infusion model is capable of providing several "long-term" progressive changes including: 1) a progressive decrease in striatal DA and HVA levels, 2) a progressive atrophy of the SN, and 3) a progressive increase in apomorphine-induced rotational behavior (Sengstock et al., 1993b). Despite numerous animal studies characterizing the effects of 6-OH DA and MPTP administration, no such progressive effects have been demonstrated to occur long-term following treatment with either of these neurotoxins. Nonetheless, one important limitation of the intranigral iron infusion model as it currently exists is that a single nigral exposure to excess iron does not induce the progressive loss of NS-DA neurons characteristic of PD. A more chronic exposure of the rat SNc to low level iron administration may ultimately achieve this most important aspect of PD pathogenesis.

What specific mechanism(s) might explain the degeneration of SNc neurons following intranigral iron infusion? This neurodegeneration most likely involves damage to lipids and/or DNA resulting from iron-catalyzed oxgen free radical formation. The role of iron in initiating the process of membranous lipid peroxidation through catalysis of hydroxyl free radical formation is, in fact, well established (Fig. 1; Youdim et al., 1989). This role for iron is consistent with our finding that intranigral iron infusions induce a substantial, albeit acute, increase in lipid peroxidation within the rat SN. Lipid peroxidation compromises membrane structure and function to an extent sufficient to cause neurodegeneration (Slater, 1984). Clearly, then, iron-catalyzed lipid peroxidation is a likely mechanism by which intranigrally-infused iron induces degeneration of NS-DA neurons in the SNc. An additional possible mechanism of iron-induced neurotoxicity within the SN also involves iron-catalyzed hydroxyl free radical formation, with ensuing damage to nuclear DNA – including DNA strand breaks and base damage (Halliwell and Gutteridge, 1989). Consistent with this possibility is an apparent heavy concentration of iron within the nucleus of SNc neurons acutely following intranigral iron infusion. In addition to induction of oxidative damage to nuclear DNA in SNc neurons, intranigral iron infusions may similarly damage mitochondrial DNA in SNc neurons. In this context, mitochondrial DNA encodes 7 of the 25 polypeptides comprising Complex I of the mitochondrial electron transport chain. Iron-catalyzed oxidative damage to mitochondrial DNA could, therefore, result in decreased Complex I activity, resulting in a depletion of cellular ATP levels and consequent neuronal death. Supportive of this premise are studies indicating that Complex I activity in the SN of PD brains is significantly reduced (Schapira et al., 1990) and that iron-loaded rat PC12 cells have reduced activities of both Complex I and IV (Hartley et al., 1993). In view of these findings, effects of intranigral iron infusion on Complex I activity in the rat SN clearly warrant investigation.

Finally, it should be mentioned that the general susceptibility of brain tissue to excess iron has been demonstrated by several studies involving the neocortical infusion of ferrous chloride (usually 28 µg Fe in 5 ul). Such infusions induce transient free radical formation and lipid peroxidation within the neocortex, as well as tissue necrosis and macrophage infiltration (Wilmore et al., 1983; Triggs and Wilmore, 1984). Infusion of ferric chloride (50 µg Fe in 5 µl) into the rat SN has recently been reported to result in substantial reductions in striatal dopaminergic markers (Ben-Shachar and Youdim, 1991). However, solutions of ferrous and ferric choride are acidic in nature and cannot be buffered without precipitation of the iron. Therefore, their effects might be due, at least in part, to acid-induced necrosis (Nedergaard et al., 1991). However, recent studies to this subject do not confirm this conclusion (Wesemann et al., personal communication). To mimic excess iron accumulation in any brain area, the amount of iron infused must be evaluated in terms of the amount of iron *normally* present in the brain area being infused; this, to avoid infusing iron amounts that are many orders of magnitude greater than those ever achieved in that brain area. In this regard, we have determined neurochemically (Sastry and Arendash, 1993) that the total amount of iron present within the SN of young adult rats is approximately 60–70 ng (i.e., about 15–20 ng/mg wet wt). The total amount of iron citrate that we infuse intranigrally ranges from 35 ng (single infusion of 0.63 nmol Fe) to 350 ng (two infusions of 3.15 nmol Fe each), and is usually in the range of 70–110 ng an amount that compares favorably with the total amount of iron present within the rat SN.

In summary, the intranigral iron infusion model provides several important aspects of PD not evident in either 6-OH DA or MPTP animal models including enhanced nigral lipid peroxidation and several long-term progressive changes. More detailed knowledge of excess nigral iron's effects and mechanism(s) of neurotoxicity could further establish this model as being a most viable one for idiopathic PD. If the intranigral iron infusion model is based on oxidative damage to nigral neurons, it should be of considerable value in assessing the efficacy of various antioxidant regimens with the potential to treat or prevent PD.

Acknowledgements

This work was funded by several grants from the United Parkinson Foundation and the Southern Medical Association. We are grateful to Dr. A. Dunn (LSU Medical Center, Shreveport, LA) for performing all HPLC determinations of striatal biogenic amine levels.

References

Basset P, Quesneau Y, Zwiller J (1986) Iron-induced L1210 cell growth: evidence of a transferrin-independent iron transport. Cancer Res 46: 1644–1647

Ben-Shachar D, Youdim MBH (1991) Intranigral iron injection induces behavioral and biochemical "parkinsonism" in rats. J Neurochem 57: 2133–2135

Corongiu FP, Dessi MA, Banni S, Bernardi S, Piccardi M, Del Zampo M, Corsini G (1987) MPTP fails to induce lipid peroxidation in vivo. Biochem Pharmacol 36: 2251–2253

Dexter DT, Carter CJ, Wells FR, Agid F, Agid Y, Lees A, Jenner P, Marsden CD (1989a) Basal lipid peroxidation in substantia nigra is increased in Parkinson's disease. J Neurochem 52: 381–389

Dexter DT, Wells FR, Lees AJ, Agid F, Agid Y, Jenner P, Marsden CD (1989b) Increased nigral iron content and alterations in other metal ions occurring in brain in Parkinson's disease. J Neurochem 52: 1830–1836

Dexter DT, Carayon A, Vidailhet M, Ruberg M, Agid F, Agid Y, Lees AJ, Wells FR, Jenner P, Marsden CD (1990) Decreased ferritin levels in brain in Parkinson's disease. J Neurochem 55: 16–20

Good PF, Olanow CW, Perl DP (1992) Neuromelanin-containing neurons of the substantia nigra accumulate iron and aluminum in Parkinson's disease – a LAMMA study. Brain Res 593: 343–346

Halliwell B, Gutteridge JM (1989) Free radicals in biology and medicine, 2nd edn. Oxford University Press, New York

Hartley A, Cooper JM, Schapira AHV (1993) Iron induced oxidative stress and mitochondrial dysfunction – relevance to Parkinson's disease. Brain Res (in press)

Irie S, Tavassoli M (1987) Tranferrin-mediated cellular iron uptake. Am J Med Sci 293: 103–111

Jellinger K, Paulus W, Grundke-Iqbal I, Riederer P, Youdim, MBH (1990) Brain iron and ferritin in Parkinson's and Alzheimer's diseases. J Neural Transm 2: 327–340

Jellinger K, Kienzl E, Rumpelmair G, Riederer P, Youdim MBH (1992) Iron-melanin complex in substantia nigra of Parkinsonian brains: an x-ray microanalysis. J Neurochem 59: 1168–1171

Kopin IJ, Schoenberg DG (1988) MPTP in animal models of Parkinson's disease. Mt Sinai J Med 55: 43–49

Nedergaard M, Goldman SA, Desai S, Pulsinelli WA (1991) Acid-induced death in neurons and glia. J Neurosci 11: 2489–2497

Olanow CW (1992) An introduction to the free radical hypothesis in Parkinson's disease. Ann Neurol 32: S2–S9

Perry TL, Yong VW (1986) Idiopathic Parkinson's disease, progressive supranuclear palsy and glutathione metabolism in the substantia nigra of patients. Neurosci Lett 67: 269–274

Riederer P, Sofic E, Rausch W-D, Schmidt B, Reynolds G, Jellinger K, Youdim MBH (1989) Transition metals, ferritin, glutathione, and ascorbic acid in Parkinsonian brains. J Neurochem 52: 515–520

Sastry S, Arendash GW (1993) Time-dependent changes in iron levels and associated neuronal loss within the substantia nigra following lesions within the neostriatum/globus pallidus complex (submitted)

Schapira AH, Cooper JM, Dexter D, Clark JB, Jenner P, Marsden CD (1990) Mitochondrial Complex I deficiency in Parkinson's disease. J Neurochem 54: 823–827

Sengstock GJ, Olanow CW, Dunn AJ, Arendash GW (1992) Iron induces degeneration of nigrostriatal neurons. Brain Res Bull 28: 645–649

Sengstock GJ, Olanow CW, Menzies RA, Dunn AJ, Arendash GW (1993a) Infusion of iron into the rat substantia nigra: nigral pathology and dose-dependent loss of striatal dopaminergic markers. J Neurosci Res 35: 67–82

Sengstock GJ, Dunn AJ, Olanow CW, Barone S, Arendash GW (1993b) Progressive changes in striatal dopaminergic markers, nigral volume, and rotational behavior following iron infusion into the rat substantia nigra (submitted)

Slater TF (1984) Free-radical mechanisms in tissue injury. Biochem J 222: 1–15

Sofic E, Riederer P, Heinsen H, Beckmann H, Reynolds GP, Hebenstreit G, Youdim MBH (1988) Increased iron (III) and total iron content in post mortem substantia nigra of parkinsonian brain. J Neural Transm 74: 199–205

Sofic E, Paulus W, Jellinger K, Riederer P, Youdim MBH (1991) Selective increase of iron in substantia nigra zona compacta of parkinsonian brains. J Neurochem 56: 978–982

Triggs WJ, Willmore LJ (1984) In vivo lipid peroxidation in rat brain following intracortical Fe^{++} injection. J Neurochem 42: 976–980

Willmore LJ, Hiramatsu M, Kochi H, Mori A (1983) Formation of superoxide radicals after $FeCl_3$ injection into rat isocortex. Brain Res 277: 393–396

Youdim MBH, Ben-Shachar D, Riederer P (1989) Is Parkinson's disease a progressive siderosis of substantia nigra resulting in iron and melanin induced neurodegeneration? Acta Neurol Scand 126: 47–54

Zigmond MJ, Abercrombie ED, Berger TW, Grace AA, Stricker EM (1990) Compensations after lesions of central dopaminergic neurons: some clinical and basic implications. TINS 13: 290–296

Correspondence: Dr. G. W. Arendash, Department of Biology and Institute for Biomolecular Science, University of South Florida, Tampa, FL 33620, U.S.A.

Iron storage and transport markers in Parkinson's disease and MPTP-treated mice

D. C. Mash[1,2], J. Singer[1], J. Pablo[1], M. Basile[1], J. Bruce[3],
and W. J. Weiner[1]

Departments of [1] Neurology, [2] Molecular and Cellular Pharmacology, and
[3] Pathology, University of Miami School of Medicine, Miami, Florida, U.S.A.

Summary

The regulation of neuronal iron is necessary for the synthesis of iron containing cytochromes and to prevent damage from free radicals by iron-oxygen interactions. We have shown that transferrin receptors are elevated over the substantia nigra in the human and rat brain (Mash et al., 1990) and are depleted concomitantly with dopaminergic terminals in the MPTP-treated mouse striatum (Mash et al., 1991). Given the iron dependency for both synthetic and degradative enzyme activities, dopaminergic neurons may express transferrin receptors on their cell surface to facilitate the uptake of iron bound to transferrin. If the intracellular iron pool is regulated by receptor-mediated transferrin uptake, then an up-regulation of transferrin receptor number may play a role in the pathogenesis of nigral cell damage in Parkinson's disease. Early in the disease process, surviving dopaminergic neurons may increase the number of transferrin receptors in order to meet the increased metabolic demand associated with compensatory changes in dopamine synthesis and turnover. The uptake of ferrotransferrin by dopaminergic neurons may result in a progressive elevation in the cellular iron load that exceeds the regulatory capacity for increased ferritin expression in the aging brain.

Introduction

Parkinson's disease is a neurodegenerative disorder that is characterized by the progressive siderosis of the substantia nigra and the marked loss of dopaminergic neurons. While the pathogenetic event that leads to the death of nigral dopaminergic neurons is unknown, many studies have focused on the role of environmental toxin and free radical damage (Kopin and Markey, 1988). The demonstration of selective increases in iron and lipid peroxidation and the corresponding decreases in glutathione oxidizing capacity within the substantia nigra in Parkinson's disease indicate that dopaminergic neurons are highly vulnerable to oxidative stress (Youdim et al., 1990; Götz et al., 1990). This cascading oxidative injury may proceed well into the end-stages of the disease (for review, Olanow, 1992). The cytotoxic events ongoing in the striatum and the substantia nigra may be stimulated by abnormal iron handling, since Fe^{2+} is known to promote oxygen radical formation (Halliwell and Gutteridge, 1986; Halliwell, 1987). Despite the detailed information on iron regulation in the periphery, there has been only recent interest in iron homeostasis in brain. At present, it is unclear whether specific alterations in iron regulatory proteins contribute to the decompartmentalization and deposition of iron in Parkinson's disease and other related movement disorders. Knowledge of the distribution and regulation of iron storage and transport proteins may be of considerable relevance to understanding the regional patterns of vulnerability in basal ganglia disorders.

Iron and other trace metals in neurodegenerative diseases

Trace metals have been linked to a diverse number of physiological functions in brain, including neurotransmitter synthesis, release, storage, and binding to dopaminergic receptor recognition sites (Dreosti and Smith, 1983; Donaldson and Barbeau, 1985). Several reports have suggested a role for trace metals in the breakdown of neuronal integrity known to occur with normal aging and in neurodegenerative diseases. Alterations in copper for Wilson's disease (Sternlieb, 1984), aluminum for Alzheimer's disease (Birchall and Chappell, 1988), and manganese (Barbeau, 1984) and iron (Dexter et al., 1987; Youdim et al., 1989; Hirsch et al., 1991) for Parkinson's disease have provided support for the hypothesis that trace element derangements in brain may cause membrane damage leading to cell death.

Regional brain iron concentrations vary widely in normal and diseased brain (Duguid et al., 1986; Rutledge et al., 1987; Norfray et al., 1988; Tennison et al., 1988; Wang et al., 1989). In pathological conditions, increased iron deposition is often highest precisely within those brain areas that have normally high iron concentrations. In the human brain, iron content is highest within the extrapyramidal system (Hock et al., 1975; Dwork et al., 1988; Dexter et al., 1989). Normal brain contains soluble iron that is visualized by Perl's staining in the globus pallidus, putamen, substantia nigra, red nucleus and subthalamic nucleus. Iron is stored primarily in an inactive form bound to intracellular ferritin. Destruction of iron-rich regions of brain leads to the appearance of granular iron in macrophages and a corresponding progressive "siderosis" of the region (Koeppen et al., 1992). The strong iron staining visable with the Perl reaction in the substantia nigra and globus pallidus in Hallervorden Spatz disease is an example of a pathological deposition of brain iron that is associated with hemosiderin or what might be more appropriately termed "siderin". Isolation of brain hemosiderin has shown that it is a degradation product of the iron storage protein ferritin (Koeppen and Dentinger, 1988). It is not known whether the deposition of hemosiderin in regions with selective neuronal cell loss is a late pathological marker of an earlier defect in brain iron storage or transport.

Brain iron storage and transport

Transferrin is an 80,000 molecular weight glycoprotein that functions as an essential growth factor to mobilize and deliver iron to the cell (Huebers and Finch, 1987). Cellular iron metabolism is self-regulated through iron dependent changes in the abundance of the iron storage protein ferritin (Theil, 1990). Transferrin measurements in rat choroid plexus demonstrate comparable rates of synthesis to those measured in the liver, which is the principal site of plasma transferrin production (Dickson et al., 1985; Aldred et al., 1987). The translocation of iron across the blood brain barrier is mediated by specific transferrin receptors located on brain microvasculature (Jeffries et al., 1984). How iron redistributes and concentrates in particular brain areas after it crosses the blood barrier is not clear. Immunocytochemical localization studies of monoclonal antibodies to rat and human transferrin have demonstrated transferrin-immunopositive oligodendrocytes, astrocytes and neurons (Swaiman and Machen, 1986; Connor and Benkovic, 1992). It is interesting to point out that transferrin will bind with significant affinity to metals other than iron, including aluminum, zinc and manganese, further suggesting a relationship of this iron transport protein with the neurotoxic actions of a number of different trace metals (Roskams and Conner, 1990).

Studies aimed at defining a particular role for ferritin, transferrin and/ or its receptors in the maintenance of brain iron homeostasis have received only minor attention. Recent studies using immunocytochemical and molecular biological techniques have demonstrated a number of important findings (for review, Connor and Benkovic, 1992). Histological studies have indicated that neuroglia play a primary role in the maintenance of brain iron. Oligodendrocytes express transferrin mRNA, and microglia and oligodendrocytes are intensely ferritin immunoreactive (Connor and Benkovic, 1992; Connor et al., 1990). In the basal ganglia, transferrin, iron and ferritin marked cells are abundant and the distribution and morphology of the cells is reportedly similar. Ferritin-positive cells are often found in the periphery of white matter tracts in the basal ganglia and subtantia nigra. Colocalization studies are needed to precisely define the cellular identity of transferrin and ferritin immunolabeled cell types. However, it is clear that in normal brain oligodendrocytes and microglia contain stainable iron and iron regulatory proteins (Connor and Benkovic, 1992). Storage of iron in a glial cell pool in areas of low neuronal densities and metabolism, such as the globus pallidus or in white matter tracts may provide a safe area with iron removed from neuronal sites that would be highly vulnerable to free radical damage (Morris et al., 1992). In keeping with this hypothesis, transferrin levels have been shown to be the highest in rat and mouse brain regions that have relatively more myelin (Connor et al., 1987). Taken together, these observations suggest that the neuroglia play an important role in the sequestration/detoxification of brain iron.

Transferrin receptor densities in Parkinson's disease

The presence of transferrin receptors in brain has been demonstrated with ligand binding assays and immunocytochemistry (Jeffries et al., 1984; Hill et al., 1985; Mash et al., 1991; Roskams and Connor, 1990). The autoradiographic localization of transferrin receptors in the rat demonstrates high densities of receptors in the cerebral cortex, hippocampus and cerebellum, with moderate densities in the caudate-putamen (Mash et al., 1990). We have previously speculated that transferrin receptor densities may be elevated over brain areas that have high metabolic requirements. Low transferrin receptors densities were visualized in the rat over the globus pallidus, a region known to have high ferritin concentrations (Hill et al., 1988). Thus, "'mismatches" in some brain areas exist between the storage of iron in ferritin and the mobilized iron required for oxidative metabolism. The regional distribution of transferrin receptors appears to correlate with the mitochondrial enzyme cytochrome oxidase (Morris et al., 1990). This pattern suggests that elevated regional densities of transferrin receptors may be correlated with a functional transport pool of iron.

It has been suggested that neuronal cell groups which have the highest transferrin receptor densities are among those most vulnerable in

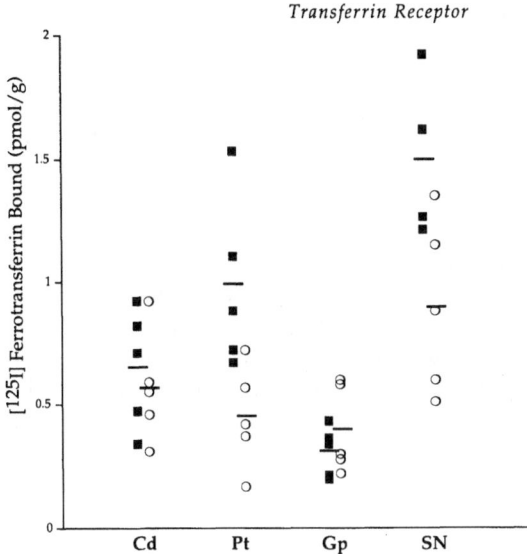

Fig. 1. Individual data from the caudate (*Cd*), putamen (*Pt*), globus pallidus (*Gp*) and the substantia nigra (*SN*) of control subjects (filled squares) and patients with Parkinson's disease (open circles) showing the density of [^{125}I]transferrin binding sites. Transferrin receptors were assayed with [^{125}I]transferrin as described previously (Mash et al., 1991). The binding site density assayed in the putamen and substantia nigra from parkinsonian patients was significantly different from age-matched control subjects ($p < 0.05$, Students t-test)

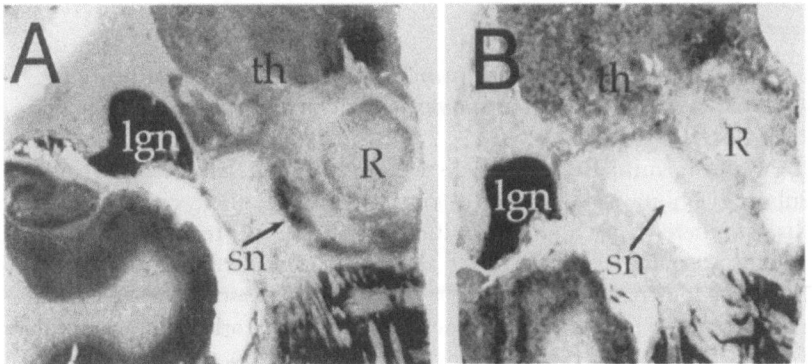

Fig. 2. Representative film autoradiograms showing the distribution of transferrin receptors in a non-neurological control subject (**A**) and in a patient with Parkinson's disease (**B**). Transferrin receptor autoradiograms were obtained as described previously (Mash et al., 1990). Note the marked loss of transferrin receptors over the substantia nigra in the parkinsonian. *lgn* lateral geniculate nucleus; *R* red nucleus; *sn* substantia nigra; *th* thalamus

neurodegenerative diseases (Morris et al., 1992). We have assayed the number of transferrin receptors in Parkinson's disease and non-neurological (control) subjects in the basal ganglia and substantia nigra. Figure 1 shows that in Parkinson's disease, the density of transferrin receptors assayed in the putamen and substantia nigra was reduced significantly as compared to age-matched subjects. Within the striatum, the pattern of dopamine loss in the Parkinsonian brain is known to be most severe in the putamen (Kopin and Markey, 1988). Previous studies in the primate have demonstrated that the putamen was more responsive to the effects of MPTP, manifesting a greater deficit in dopaminergic markers than in the caudate nuclei (Irwin et al., 1989). In normal subjects, the density of transferrin receptors was highest in the substantia nigra and putamen, with lower numbers of sites assayed in the globus pallidus and the caudate nucleus (Fig. 1). Transferrin receptor densities in the putamen and substantia nigra were reduced significantly in idiopathic Parkinson's disease (p > 0.05). The reduction in the total number of transferrin receptors assayed in the Parkinson's disease subjects ranged from 20 to 50% of control values. In contrast, the density of transferrin receptors assayed in the head of the caudate nucleus was not significantly different from values obtained in age-matched control subjects. These data are in keeping with previous neurochemical observations which demonstrate that in idiopathic Parkinson's disease, the dopamine loss in the putamen was more severe than in the caudate nucleus (Kish et al., 1988). The significant reduction of transferrin receptors in the substantia nigra in Parkinson's disease suggests that transferrin receptors may be expressed by nigral dopaminergic neurons (Fig. 2).

Ferritin mRNA regulation in Parkinson's disease and other related movement disorders

While in mild Parkinson's disease iron content is unchanged in the substantia nigra as compared to control subjects, in severe Parkinson's disease there is a highly significant increase in total iron content (Youdim et al., 1989; Dexter et al., 1992). The potential toxicity of a progressive increase in iron load in Parkinson's disease would be dependent upon the intracellular stores of ferritin and other low molecular weight iron binding proteins (Dexter et al., 1992). Dexter and coworkers (1992) have demonstrated that the increases in iron were not associated with elevated levels of ferritin protein in Parkinson's disease. In contrast, regional assays in Parkinson-plus syndromes (multiple system atrophy and Progressive Supranuclear Palsy) demonstrated increased or normal amounts of ferritin. These results suggest that in Parkinson's disease the iron load in the substantia nigra may exceed the available storage capacity of ferritin.

We have used a riboprobe specific to the ferritin heavy chain to measure the relative levels of ferritin mRNA by Northern blot and in situ hybridization histochemistry. In Parkinson's disease, there was a decrease in ferritin mRNA in the substantia nigra compared to age-matched control subjects (Figs. 3 and 4). Our preliminary analysis showed that the ferritin hybridization signal was markedly elevated in Parkinsonian brain over surviving dopaminergic neurons (data not

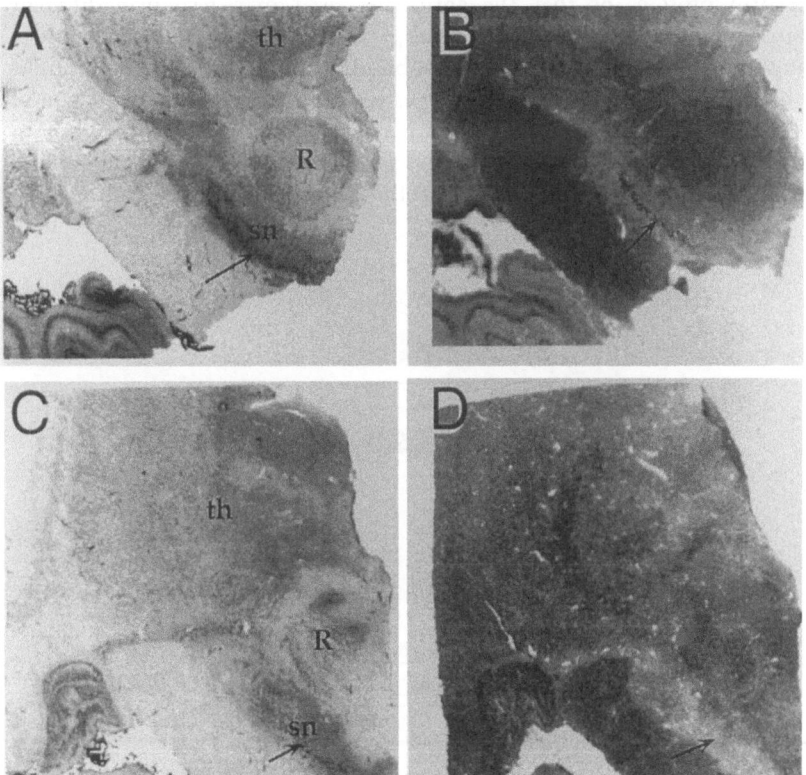

Fig. 3. Distribution of ferritin heavy chain mRNA in an age-matched control subject (A, B) and a patient with Parkinson's disease (C, D). Panels B and D are the film autoradiograms and panels A and C show the Nissl stained sections for locations of cytoarchitectural landmarks. Cryostat sections were hybridized with a 35S-labeled RNA probe specific for the human ferritin heavy chain gene. Note the marked decrease in the levels of ferritin mRNA over the substantia nigra (*Sn*) in a patient with Parkinson's disease (D). *th* thalamus, *R* red nucleus

shown). Northern blot analysis of regional brain RNA demonstrated significant reductions in ferritin expression in the putamen and substantia nigra in Parkinson's disease (Fig. 4). In keeping with the previous report on the level of ferritin protein by Dexter and colleagues (1992), we observed normal or elevated expression of ferritin mRNA in Parkinson-plus syndromes (Fig. 4). Taken together, these findings provide support for a generalized decrease in ferritin synthesis in Parkinson's disease.

These observations are in contrast to a previous report by Riederer and coworkers (1989), who demonstrated an increase in ferritin levels in the substantia nigra in Parkinsonian brain. It has been suggested previously, that this discrepancy may be due to different techniques for measuring ferritin protein, or that the antibodies used may recognize different ferritin heteropolymers (Dexter et al., 1992). Ferritin is a protein consisting of two partially homologous subunits, the heavy (H) and the light chains (L). Tissue ferritin phenotypes are regulated by the expression of these two subunits, with the proportion of each subunit varying in different tissues. In addition to its iron detoxification and storage functions, ferritin may play a role in cellular iron flux (Boyd et al., 1985).

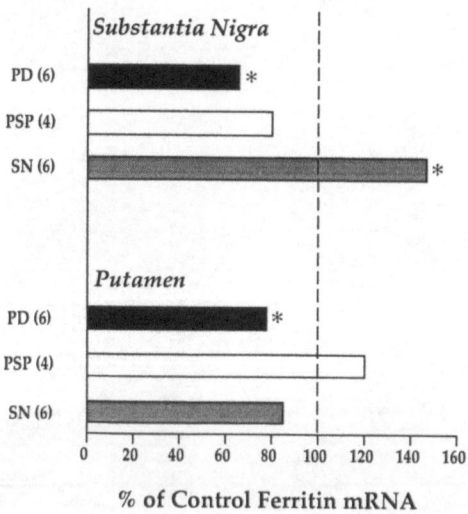

Fig. 4. Ferritin heavy chain mRNA levels in Parkinson's disease (*PD*), Striatonigral Degeneration (*SN*), and Progressive Supranuclear Palsy (*PSP*). Dot blot analysis was performed on total RNA extracted from the substantia nigra and putamen and probed with a ^{32}P-labeled riboprobe complementary to the human ferritin heavy chain gene. Values were normalized to an internal actin standard and are presented as the percentage of control values. The number of subjects matched for age and autolysis times in each group is indicated in parenthesis

Boyd and coworkers (1985) have suggested that human ferritins functionally diverged during evolution into H-ferritin ("uptake and release") and L-ferritin ("storage"). Changes in ferritin subunits have been described during fetal development, with increased heavy subunits occuring early in gestation and converting to light subunits during the third trimester (Drydale and Alpert, 1975). Such developmental changes may be relevant to the aging brain, for which iron and other trace metals requirements may vary considerably over the lifespan (Joshi and Zimmerman, 1988). One possible explanation for the discrepancy noted above is that neurons may have an increased complement of H-ferritin, while microglia and oligodendrocytes have elevated expression of L-ferritin. If there are differences in the ratios of neuronal and glia ferritin H- and L-subunit expression, than the decrease in H-ferritin should parallel more closely the severity of dopaminergic neuronal loss in Parkinson's disease. Studies are in progress to measure the relative abundance of H- and L-ferritin mRNAs in Parkinson's disease and other basal ganglia disorders. Additional studies are needed to address the specific cell type using antisera directed against the two ferritin forms with immunohistochemical colocalization techniques.

Effects of MPTP treatment on transferrin receptor and ferritin expression

We have assessed temporal changes in iron storage and transport markers in the MPTP-treated mouse model of Parkinson's disease. We have demonstrated a significant reduction in the density of both transferrin receptors and [³H]-mazindol binding to dopamine transporters in the mouse striatum 7 days post MPTP treatment (Mash et al., 1991). The time course for the reduction in the expression of transferrin receptor mRNA preceded the reduction in the number of binding sites assayed with [¹²⁵I]-transferrin (Fig. 5). We did not observe a marked change in the abundance of H-ferritin mRNA until 10 days post MPTP treatment. Quantitative densitometric analysis of [¹²⁵I]-ferrotransferrin and [³H]-mazindol binding site autoradiograms demonstrated a rapid recovery in [¹²⁵I]-ferrotransferrin binding in the striata of mice sacrificed at later survival times (Mash et al., 1991). The trend toward recovery of striatal [³H]-mazindol binding lagged temporally behind the "normalized" expression of transferrin receptors.

In MPTP-treated C57 black mice, dopaminergic parameters are known to show variable rates of recovery depending upon the dose regimen of MPTP (Ricaurte et al., 1986; Sonsalla and Heikkila, 1986; Ricaurte et al., 1987; Reinhard et al., 1988). It has been suggested that the recovery in dopaminergic markers following MPTP treatment may result from the collateral sprouting of undamaged dopaminergic terminals (Irwin et al., 1989; Kitt et al., 1986; Sonsalla and Heikkila, 1986; Willis and Donnan, 1987). In addition, compensatory changes in dopamine metabolism may account, in part, for the neurochemical recovery observed in some studies (Irwin et al., 1989; Sonsalla and Heikkila, 1986). The sustained recovery of transferrin binding observed in the striatum at the later time points may be due to the regenerative sprouting of dopaminergic afferents. However, we have speculated that the apparent "early" recovery in transferrin receptor densities may result from an up-regulation of transferrin receptor numbers in surviving striatal dopaminergic terminals as a compensatory cellular adaptation (Mash et al., 1991). Early compensatory changes in dopaminergic transmission

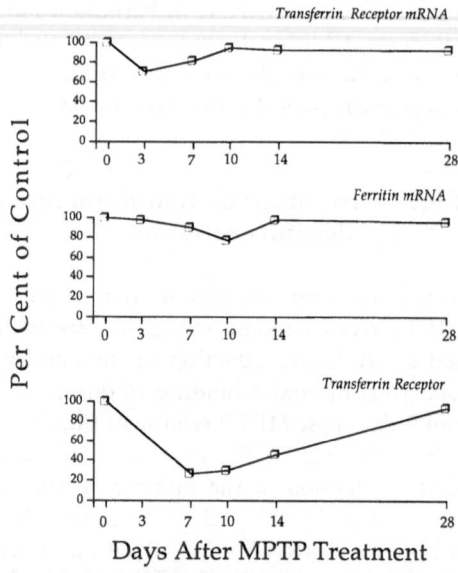

Fig. 5. Time course for the effects of MPTP treatment on the density of transferrin receptor and the levels of ferritin and transferrin receptor mRNAs. Transferrin receptor numbers were obtained from region-of-interest densitometric analysis in the striatum from film autoradiograms (Mash et al., 1991). For measures of the relative abundance of transferrin receptor and ferritin mRNAs, total RNA was isolated from dissected mouse brains (5 pooled diencephalon to mesencephalon blocks), fractionated on a formaldehyde agarose gel and transferred to nitrocellulose. The RNA isolated from MPTP-treated mice at different survival times was hybridized to a ^{32}P-labeled antisense RNA probe to a ferritin heavy chain genomic clone and to a transferrin receptor cDNA clone. Quantitative analysis of the northern blots was carried out by scanning the intensity of the bands on the autoradiographic film using the Howtek Scanmaster 3 and Brain (1.6) software

after exposure to MPTP may lead to increased expression of transferrin receptors by either surviving dopaminergic neurons and/or input-sensitive target neurons to meet the increased metabolic demand resulting from striatal dopaminergic denervation. Studies are in progress to visualize the regional expression and time course for ferritin and transferrin receptor mRNAs, to further determine the coordinated synthesis of these two metabolically-related proteins following MPTP exposure.

Acknowledgements

This work was supported by a grant from the National Parkinson Foundation, Inc., Miami, FL.

References

Aldred A, Dickson P, Marley P, Schreiber G (1987) Distribution of transferrin synthesis in brain and other tissues of the rat. J Biol Chem 262: 5293–5297

Barbeau A (1984) Manganese and extrapyramidal disorders a critical review and tribute to Cotzias, George C. Neurotoxicology 5: 13–35

Birchall J, Chappell J (1988) Aluminum, chemical physiology and Alzheimer's disease. Lancet ii: 1008–1010

Boyd D, Vecoli C, Belcher D, Jain S, Drysdale J (1985) Structural and functional relationships of human ferritin H and L chains deduced from cDNA clones. J Biol Chem 260: 11755–11761

Connor J, Phillips T, Lakshman M, Barron K, Fine R, Csiza C (1987) Regional variation in the levels of transferrin in the CNS of normal and myelin deficient rats. J Neurochem 49: 1523–1529

Conner J, Menzies S, St. Martin S, Mufson E (1990) Cellular distribution of transferrin, ferritin and iron in normal and aged human brains. J Neurosci Res 27: 595–611

Conner J, Benkovic S (1992) Iron regulation in the brain: histochemical, biochemical and molecular considerations. Ann Neurol 32: 551–561

Dexter D, Wells F, Lees A (1987) Increased nigral iron content in post-mortem parkinsonian brain. Lancet ii: 1219-1220

Dexter D, Carter F, Javoy-Agid F, Agid Y, Lees A, Jenner P, Marsden C (1989) Increased nigral iron content and alterations in other metal ions occuring in brain in Parkinson's disease. J Neurochem 52: 1830–1836

Dexter D, Jenner P, Schapira A, Marsden C (1992) Alterations in levels of iron, ferritin, and other trace metals in neurodegenerative diseases affecting the basal ganglia. Ann Neurol 32: S94–S100

Dickson P, Aldred A, Marley P, Guo-Fen T, Howlett G, Schreiber G (1985) High prealbumin and transferrin mRNA levels in the choroid plexus of rat brain. Biochem Biophys Res Commun 127: 890–895

Donaldson J, Barbeau A (1985) Manganese neurotoxicity: possible clues to the etiology of human brain disorders. In: Neurology and neurobiology, vol 15. Metal ions in neurology and psychiatry. Liss, New York, pp 259–285

Dreosti I, Smith R (1983) Neurobiology of trace elements, vol 1. Humana, Clifton NJ, pp 269–291

Drysdale J, Alpert E (1975) Human isoferritins - reply. Br J Haem 30: 518-519

Duguid J, De La Paz R, DeGroot J (1986) Magnetic resonance imaging of the midbrain in Parkinson's disease. Ann Neurol 20: 744–747

Dwork A, Schon E, Herbert J (1988) Nonidentical distribution of transferrin and ferric iron in human brain. Neuroscience 27: 333–345

Götz M, Freyberger A, Riederer P (1990) Oxidative stress: a role in the pathogenesis of Parkinson's disease. J Neural Transm [Suppl] 29: 241–249

Halliwell B (1987) Oxidants and human disease: some new concepts. Fed Am Soc Exp Biol 892: 358–364

Halliwell B, Gutteridge J (1986) Oxygen free radicals and iron in relation to biology and medicine: some problems and concepts. Mol Aspects Med 8: 89–193

Hill J (1988) Brain iron: neurochemical and behavioral aspects. Taylor and Francis, London, pp 1–24

Hill J, Ruff M, Weber R, Pert C (1985) Transferrin receptors in rat brain: neuropeptide-like pattern and relationship to iron. Proc Natl Acad Sci USA 82: 4553–4557

Hirsch E, Brandel J, Galle P, Javoy-Agid F, Agid Y (1991) Iron and aluminum increase in the substantia nigra of patients with Parkinson's disease: an x-ray microanalysis. J Neurochem 56: 446–451

Hock A, Demmel U, Schicha H, Kasperek K, Feinendegen L (1975) Trace element concentration in human brain. Brain 98: 49-64

Huebers H, Finch C (1987) The physiology of transferrin and transferrin receptors. Physiol Rev 67: 520-582

Irwin I, DeLaney L, Forno L, Finnegan K, Di Monte D, Langston J (1989) The evolution of nigrostriatal neurochemical changes in the MPTP-treated squirrel monkey. Brain Res 531: 242-252

Jefferies W, Brandon M, Hunt 5, Williams A, Gatters K, Masons D (1984) Transferrin receptor on endothelium of brain capillaries. Nature 312: 162–163

Joshi J, Zimmerman A (1988) Ferritin: an expanded role in metabolic regulation. Toxicology 48: 21-29

Kish S, Shannak K, Hornykiewicz O (1988) Uneven pattern of dopamine loss in the striatum of patients with idiopathic Parkinson's disease: pathophysiological and clinical implications. N Engl J Med 318: 876–880

Kitt C, Cork L, Eidelberg F, John T, Price D (1986) Injury of nigral neurons exposed to 1-methyl-4-phenyl 1,2,3,6-tetrahydropyridine: a tyrosine hydroxylase immunocytochemical study in monkey. Neuroscience 17: 1089–1103

Koeppen A, Dentinger M (1988) Brain hemosiderin and superficial siderosis of the central nervous system. J Neuropathol Exp Neurol 47: 249–270

Koeppen A, Hurwitz C, Dearborn R, Dickson A, Borke R, Chu R (1992) Experimental superficial siderosis of the central nervous system: biochemical correlates. J Neurol Sci 112: 38-45

Kopin I, Markey S (1988) MPTP toxicity: implications for research in Parkinson's disease. Ann Rev Neurosci 11: 81-96

Mash D, Pablo J, Flynn D, Efange S, Weiner W (1990) Characterization and localization of transferrin receptors in the rat brain. J Neurochem 55: 1972-1979

Mash D, Pablo J, Buck B, Sanchez-Ramos J, Weiner W (1991) Distribution and number of transferrin receptors in Parkinson's disease and in MPTP-treated mice. Exp Neurol 114: 73-81

Morris C, Candy J, Bloxham C, Edwardson J (1990) Brain transferrin receptors and the distribution of cytochrome oxidase. Biochem Soc Trans 18: 647

Morris C, Candy J, Keith A, Oakley A, Taylor G, Pullen G, Bloxham C, Gocht A, Edwardson J (1992) Brain iron homeostasis. J Inorg Biochem 47: 257-265

Norfray J, Couch J, Elble R, Good D, Manyam B, Patrick J (1988) Visualization of brain iron by mid-field MR. Am J Neurorad 9: 77-82

Olanow C (1992) An introduction to the free radical hypothesis in Parkinson's disease. Ann Neurol 32: 52-59

Reinhard J, Miller D, O'Callaghan J (1988) The neurotoxicant MPTP (1-methyl-4-phenyl-1,2,3,6-tetrahydropyridine) increases glial fibrillary acidic protein and decreases dopamine levels of the striatum: evidence for glial response to injury. Neurosci Lett 95: 246-251

Ricaurte G, Langston J, Delanney L, Irwin I, Peroutka S, Forno L (1986) Fate of nigrostriatal neurons in young mature mice given 1-methyl-4-phenyl-1, 2, 3, 6-tetrahydropyridine: a neurochemical and morphological reassessment. Brain Res 375: 117-124

Ricaurte G, Delanney L, Irwin I, Langston J (1987) Older dopaminergic neurons do not recover from the effects of MPTP. Neuropharmacology 26: 97-99

Riederer P, Sofic E, Rausch W, Schmidt B, Reynolds G, Jellinger K, Youdim M (1989) Transitional metals, ferritin, glutathione, and ascorbic acid in parkinsonian brains. J Neurochem 52:515-520

Roskams A, Connor J (1990) Aluminum access to the brain: a possible role for the transferrin receptor. Proc Natl Acad Sci USA 87: 9024-9027

Rutledge J, Hilal S, Silver A, Defendini R, Fahn S (1987) Study of movement disorders and brain iron by MR. Am J Neurorad 8: 397-411

Sonsalla P, Heikkila R (1986) The influence of dose and dosing interval on MPTP-induced dopaminergic neurotoxicity in mice. Eur J Pharmacol 129: 339-345

Sternlieb I (1984) Wilson's disease: indication for liver transplants. Hepatology 4: 515-517

Swaiman K, Machen V (1986) Iron uptake by mammalian cortical neurons. Ann Neurol 16: 6670

Tennison M, Bouldin M, Whaley R (1988) Mineralization of the basal ganglia detected by CT in Hallervorden-Spatz syndrome. Neurology 38: 154-155

Theil E (1990) Regulation of ferritin and transferrin receptor mRNAs. J Biol Chem 265 (9): 4771-4774

Wang J, Huang C, Hwang Y, Chiang J, Lin J, Chen J (1989) Manganese induced parkinsonism: an outbreak due to an unrepaired ventilation control system in a ferromanganese smelter. Br J Ind Med 46: 856-859

Willis G, Donnan G (1987) Histochemical, biochemical and behavioral consequences of MPTP treatment in C-57 black mice. Brain Res 402: 269–274
Youdim M, Ben-Shachar D, Riederer P (1989) Is Parkinson's disease a progressive
siderosis of substantia nigra resulting in iron and melanin induced neurodegeneration? Acta Neurol Scand 126: 47–54
Youdim M, Ben-Shachar D, Yehuda Y, Riederer P (1990) The role of iron in the
basal ganglion. Adv Neurol 53: 155–161

Correspondence: Dr. D.C. Mash, Department of Neurology (D4–5), University of
Miami School of Medicine, 1501 N.W. 9 Avenue, Miami, FL 33141, U.S.A.

Pathogenesis of Parkinson's disease: iron and mitochondrial DNA deletion

Y. Mizuno, H. Mochizuki, K. Nishi, S.-i. Ikebe, N. Hattori, and Y. Hattori-Nakagawa

Department of Neurology, Juntendo University School of Medicine, Japan

Summary

Numbers of biochemical abnormalities which may be relevant to the degenerative process of nigral dopaminergic neurons have been described. These include accumulation of iron in substantia nigra, decrease in the enzymatic activity and the amount of subunits of mitochondrial complex I, increase in the amount of deleted mitochondrial DNA, and possible increase in oxygen derived free radicals. Recent progress in this field is reviewed in this communication. Although the primary cause of Parkinson's disease is still unknown, these abnormalities listed above will contribute to the progression of the degenerative process. In addition, we report our recent data on the toxic effects of iron and synthetic dopamine melanin on cultured dopaminergic neurons, and discuss possible interaction of iron and mitochondrial DNA.

Iron is implicated in the pathogenesis of nigral cell death in Parkinson's disease (PD), as first demonstrated by Youdim et al. (1989) and Riederer et al. (1989) and then by other workers (Dexter et al., 1989b; Jellinger et al., 1990; Hirsch et al., 1991; Sofic et al., 1991; Ben-Shachar et al., 1991); its content is increased in substantia nigra of patients with PD. Regarding the subcellular location of iron, Jellinger et al. (1992) and Good et al. (1992) recently revealed association of Fe^{3+} with neuromelanin by X-ray microanalysis. Therefore, it seems quite probable that Fe^{3+} is playing a role in the progression of nigral degeneration. The question as to if iron is playing a primary role in initiating the degenerative process or not should be further investigated, because iron accumulation was not found in mild PD patients in one report (Riederer et al., 1989). On the other hand, numbers of mitochondrial abnormalities have been disclosed in PD as reviewed later in this communication. In this paper, we will review the recent progress in the molecular biological abnormalities which appear to have a significance in the pathogenesis of PD, and discuss possible interaction between iron and mitochondria in PD including our recent data on the effects of iron on the cultured dopaminergic neurons. First of all, I would like to review briefly what the electron transfer complexes are like.

Mitochondrial electron transfer complexes

The mitochondrial electron transfer complexes are the protein-lipid enzyme complexes located in the inner membrane of mitochondria (Hatefi, 1985 for review). The mitochondrial oxidative phosphorylation can be achieved by the cooperative actions of these complexes. They consist of five complexes, that is, NADH : ubiquinone oxidoreductase (complex I), succinate : ubiquinone oxidoreductase (complex II), ubiquinol : ferricytochrome c oxidoreductase (complex III), ferrocytochrome c : oxygen oxidoreductase (complex IV), and ATP synthase (complex V). As the electrons are transferred from NADH or succinate to molecular oxygen, protons are translocated from the matrix side to the intermembrane space, and this protonic energy is utilized for ATP synthesis by complex V. Each complex has a finite number of subunits (Table 1) which are proteins with or without transition metals or quinoid compounds, or proteins associated with phospholipids. The actual electron carriers of the respiratory chain include quinoid compounds and transition-metal complexes as shown in Table 2.

Complex I is the major entry point for electrons into the respiratory chain. Complex I removes electrons from NADH and passes them via a series of enzyme-bound redox centers to ubiquinone. Four protons are considered to be pumped out by complex I per electron pair transferred from NADH to ubiquinone (Weiss and Friedrich, 1991). The number of subunits of complex I isolated from bovine heart mitochondria was determined by 1- and 2-dimensional polyacrylamide gel electrophoresis, and it was concluded that the bovine complex contained about 25 different subunits (Hatefi, 1985; Ragan, 1987 for review). But recent protein sequence analysis revealed that complex I was composed of at least 41 subunits containing more than 7724 amino acids of unique sequences (Walker et al., 1992; Walker, 1992 for review). Seven of 41 subunits are encoded by the mitochondrial DNA (mtDNA) (Anderson et al., 1981; Chomyn et al., 1985, 1986). All seven of them are hydrophobic intrinsic membrane proteins.

Complex I can be divided into three distinct fractions which are

Table 1. The numbers of subunits composing the mammalian electron transfer complexes

	Total	Nuclear coded	Mitochondrially coded
Complex I	41	34	7
Complex II	4	4	0
Complex III	10	9	1
Complex IV	13	10	3
Complex V	18	16	2

(Merle and Kadenbach, 1980; Capaldi, 1982; Hatefi, 1985; Walker, 1992)

Table 2. Electron carriers of the respiratory chain

Complex I	FMN, Iron-sulfur clusters, Ubiquinone
Complex II	FAD, Iron-sulfur clusters, Ubiquinone
Complex III	Cyt b_{562}, Cyt b_{566}, Cyt c_1, Iron-sulfur clusters
Complex IV	Cyt $aCua$, Cyt a_3Cua_3

(Hatefi, 1985)

referred to as flavoprotein fraction, iron-sulfur protein fraction, and hydrophobic fraction (Hatefi, 1985 for review). The flavoprotein fraction is water-soluble, and consists of three proteins (51 kDa, 24 kDa, and 10 kDa) and one molecule of flavin-mononucleotide (FMN) per complex I. The flavoprotein fraction retains an NADH-dehydrogenase activity, and NADH-binding site is considered to lie within the 51 kDa subunit. The iron-sulfur fraction is water-soluble, and consists of 7 proteins referred to as 75 kDa, 49 kDa, 30 kDa, 18 kDa, 15 kDa, 13 kDa and a subunit called B13 (Walker, 1992 for review). The hydrophobic fraction contains many water-insoluble proteins including all 7 mito-chondrially encoded subunits constituting membrane proteins, however, it also contains some water-soluble globular proteins, some of which also contain iron-sulfur clusters. For instance, the subunits called TYKY according to the Walker's nomenclature has two [4F-4S] clusters, and PSST contains one [4F-4S] cluster (Walker, 1992 for review). The exact chemical natures of the water-insoluble fraction have not been well elucidated yet.

The bovine complex I is considered to contain a total of approximately 24 atoms of iron per complex, and iron exists in the form of either [4Fe-4S] or [2Fe-2S] clusters (Ragan et al., 1982; Ragan, 1987 for review). The iron-sulfur clusters serve as electron carriers. The 51 kDa subunit of the flavoprotein fraction contains one [4Fe-4S] cluster per subunit, and the 24 kDa subunits one [2Fe-2S] cluster per subunit. The 75-kDa subunit of iron-sulfur protein fraction is considered to contain one [4Fe-4S] and one [2Fe-2S] clusters, and the TYKY and the PSST subunits, which are usually associated with the hydrophobic fraction of complex I, contain two [4Fe-4S] clusters and one [4Fe-4S] cluster, respectively (Walker, 1992 for review).

Well known specific inhibitors of complex I, rotenone, piericidin A, and barbiturates, are considered to act in the same region near the ubiquinone-binding site inhibiting the electron transfer from Fe-S cluster N-2 to ubiquinone (Walker, 1992). The parkinsonism-inducing neuro-toxin, 1-methyl-4-phenylpyridinium ion (MPP$^+$), also binds to the same site (Singer and Ramsay, 1990). There is still a great uncertainty about the actual electron pathway of complex I.

Complex II removes electrons from succinate, and passes them to ubiquinone. Mammalian complex II is composed of four proteins all encoded by nuclear DNA. Succinate dehydrogenase is the major compo-nent of complex II, and it contains one mole of covalently bound flavin-adenine dinucleotide per complex (Hatefi, 1985 for review).

Complex III removes electrons from ubiquinol, and passes them to ferricytochrome c. This reaction is coupled to transmembrane proton translocation. The mammalian complex III is composed of 10 unlike proteins, of which core protein 1, core protein 2, cytochrome b, cyto-chrome c_1, and Fe-S protein constitute major subunits (Capaldi, 1982). Cytochrome b is encoded by mtDNA (Anderson et al., 1982), and the remaining by nuclear DNA. Three out of 10 subunits are associated with redox centers, i.e., cytochrome b_{562}-b_{566}, cytochrome c_1, and a [2Fe-2S] cluster (Hatefi, 1985).

Complex IV removes electrons from ferrocytochrome c and passes them to molecular oxygen. This reaction is coupled to transmembrane proton translocation. The mammalian complex IV is composed of 12 or 13 unlike proteins (Merle and Kadenbach, 1980; Capaldi, 1982). The three largest subunits of complex IV (subunits I, II, and III) are encoded by mtDNA (Anderson et al., 1982).

Complex V is responsible for ATP synthesis from ADP and inorganic phosphate at the expense of protonic energy derived from the operation of the respiratory complexes I, III, and IV (Hatefi, 1985). Complex V can be divided into F_1 and F_0 sectors. F_1 is concemed with ATP synthesis, and the F_0 proton translocation (Kagawa, 1979). The number of subunits of mammalian complex V is considered to be 18 (Capaldi, 1982 for review), of which subunits 6 and 8 are encoded by mtDNA (Anderson et al., 1982).

Mitochondrial abnormalities in Parkinson's disease

In PD, mitochondrial abnormalities have been studied from three related aspects, i.e., enzymatic activities of the electron transfer complexes, subunit analysis by immunoblotting and immunohistochemistry, and mtDNA analysis for deletions and point mutations. Table 3 summarizes complex I activity of PD reported in the literature.

Schapira et al. (1989, 1990a) and Reichmann et al. (1990, 1993) reported decrease in complex I activity in PD. Although we could not reproduce this decrease (Mizuno et al., 1990), we found decreases in the amounts of complex I subunits by Western blotting (Mizuno et al., 1989) and by immunohistochemistry (Hattori N et al., 1991). Immunohistochemistry revealed uneven distribution of immuno-stainable complex I from one neuron to another suggesting the presence of different energy status among the nigral melanin-containing neurons, which may be interpreted as an evidence of energy mosaicism. The energy mosaicism has been implicated as one of the most important mechanisms of neuronal death associated with ageing (Linnane et al., 1989, 1990).

In tissues other than the brain, Bindoff et al. (1989) reported decreases in the activities not only of complex I but also of complex II and IV in the skeletal muscle. Our group found a decrease in the muscular complex I activity (Nakagawa-Hattori et al., 1991), but Schapira's group found normal muscular activity (Mann et al., 1991). Shoffner et al. (1991) found a significant decrease in the activity of complex I using the biopsied skeletal muscles in 4 out of 6 patients with Parkinson's disease. Therefore, three out of four reports found a decrease in the complex I activity of the skeletal muscle, but one report did not. In our opinion, it is not surprising if the skeletal muscle showed a decrease in the complex I activity, as the skeletal muscle is a post-mitotic tissue, there may well be an accumulation of noxious stimuli not only in the brain but also in the skeletal muscle, suppose PD is caused by an exogenous or endogenous toxin.

Table 3. Mitochondrial complex I activity in Parkinson's disease

	Decreased	Equivocal	Normal
Brain	Schapira et al. (1989) Reichmann et al. (1993)	Mizuno et al. (1990)	
Skeletal muscle	Bindoff et al. (1989) Shoffner et al. (1991) Nakagawa-Hattori et al. (1992)		Mann et al. (1992)
Platelets	Parker et al. (1989)	Yoshino et al. (1992)	Mann et al. (1992) Kriege et al. (1992)

Regarding the platelet, Parker et al. (1989) reported a significant decrease in the complex I activity. We found a marginal decrease in the platelet complex I activity of questionable significance (Yoshino et al., 1991). Other reports could not reproduce Parker's data (Mann et al., 1992; Krige et al., 1992). Therefore, it seems rather unlikely that the platelet complex I is affected in PD.

As mtDNA encodes 13 subunits of the electron transfer complexes, we analyzed mtDNA for mutations in PD. We found the common 5-kb deletion encompassing the genes for ND 5 and ATPase 6/8 subunits in all the patients with PD studied by the PCR method (Ikebe et al., 1990). By a quantitative analysis, the amount of deletion was estimated to be approximately 5% of total striatal mtDNA (Ozawa et al., 1990). This deletion could not be detected by Southern blotting (Schapira et al., 1990b; Lestienne et al., 1990). The failure to detect this deletion by Southern blot analysis is probably due to the small amount of the deleted mtDNA compared to the normal mtDNA. Recently age-associated increase in deleted mtDNA in the central nervous system was reported by three groups, and the increase was most prominent in putamen (Simonetti et al., 1992; Soong et al., 1992; Corral-Debrinski et al., 1992). Therefore, it seems quite conceivable that mtDNA deletion does occur in human brains not only in ageing but also in PD. The question is if the deleted mtDNA found in PD represents a mere ageing process, or if it has a role in the degenerative process of substantia nigra.

In organs other than the brain, similar ageing-associated increase in deleted mtDNA has been reported in the heart muscle (Cortopassi and Amheim, 1990; Hattori K et al., 1991), and in the liver (Yen et al., 1991). We had a chance to analyze skeletal muscle mtDNA for deletions in PD, and both control and parkinsonian patients showed the common 5-kb deletion as they became older, but parkinsonian patients showed this deletion earlier in the age than the control subjects (unpublished observation). Therefore, the metabolic defects taking place in parkinsonian substantia nigra seem to have a similarity to the metabolic change of the ageing process. Parkinson's disease may be regarded as an example of localized accelerated ageing.

The next question relates the possible cause of mitochondrial abnormalities in PD.

What may be the cause of mitochondrial abnormalities in Parkinson's disease?

The magnitude of the decrease in complex I activity in PD is not big enough to indicate a congenital defect. Therefore, this decrease is more likely a secondary phenomenon to something else. If a toxin like MPTP (1-methyl-4-phenyl-1,2,3,6-tetrahydropyridine) is involved in the pathogenesis of PD, it is easy to understand specific decrease of complex I in this disease, as MPP$^+$ specifically inhibits complex I (Nicklas et al., 1985; Heikkila et al., 1985; Ramsay et al., 1986; Mizuno et al., 1986, 1987). Along this line numbers of chemicals resembling MPTP have been screened for toxicity against nigral cells. For instance, tetrahydroisoquinoline (TIQ) was found to produce a condition similar to parkinsonism in primates (Nagatsu and Yoshida, 1988; Yoshida et al., 1990). Furthermore, TIQ was found to be N-methylated in mammalian brains (Naoi et al., 1989a), and N-methyl-TIQ is oxidized by monoamine oxidase to MPP$^+$-like ion, N-methylisoquinolinium ion (Naoi et al., 1989b). Thus formed N-methylisoquinolinium ion inhibits the mitochondrial state 3 respiration and complex I activity like MPP$^+$ (Suzuki et al., 1992). But its toxic potency is much weaker than that of MPP$^+$, and it seems unlikely that TIQ or N-methylisoquinolinium ion is the cause of PD.

Another endogenous brain-bioactivated candidate toxin is β-carbolinium ion (Collins et al., 1992). Interestingly this compound inhibits both dopamine uptake and complex I activity like MPP$^+$ (Drucker et al., 1990; Albores et al., 1990). But none of these compounds so far have been found increased in parkinsonian brains. Presence of TIQ has also been shown in human brains (Niwa et al., 1987), but again it has not been shown that it is increased in PD. On the contrary, one of the TIQ derivatives, 1-methyl-1,2,3,4-tetrahydroisoquinoline (1-methyl-TIQ) was reported to be decreased in parkinsonian brains (Ohta et al., 1987), and 1-methyl-TIQ could prevent MPTP-induced behavioral defects in mice (Tasaki et al., 1991).

Thus neurotoxin is a very attractive hypothesis, however, so far no clear evidence exists to indicate that PD is in fact caused by a neurotoxin. Therefore, other possibilities should be sought. Another possibility is that complex I deficiency may be caused by selective accumulation of deleted mtDNA in nigral neurons. If there is an uneven distribution in the deleted mtDNA among nigral neurons, those expressing a large amount of deleted mtDNA may die earlier than those expressing a small amount of deleted mtDNA. To test this hypothesis, in situ hybridization study is necessary, which is undergoing now, but at this moment data are not enough to prove or disprove this possibility. Another question to be answered is if mtDNA deletion is a consequence of point mutations, or not.

As mtDNA is estimated to mutate 10 times the frequency of nuclear DNA due to oxidative damage (Richter et al., 1988), there is a large individual difference in the nucleotide sequence of mtDNA. Therefore, we looked for point mutations of mtDNA in PD. The reasons why mtDNA has a higher incidence of point mutations include lack of histon protein coating (Caron et al., 1979), poor repair mechanism (Clayton, 1982), absence of introns except for the very small segment (Anderson et al., 1981), a high turnover rate (Gross et al., 1969), and a high error rate of DNA polymerase-gamma (Kunkel and Loeb, 1981) which is believed to be responsible for mtDNA replication (Clayton, 1982). We sequenced whole mtDNA in 5 patients with PD, and compared those sequences with those of the Anderson's Cambridge sequence and other control patients (Ikebe et al., submitted). We found 1 to 2 nonsynonymous point mutations which were seen only infrequently in control subjects, in every patients with PD studied. But we could not find a point mutation that was common to PD. The question if these mutations represent only polymorphisms of mtDNA, or they may endorse some susceptibility for degradation to those subunits encoded in mutated mtDNA awaits further investigation. Another question to be studied is if point mutations in certain areas of mtDNA may predispose to larger deletions or not. These aspects of mitochondrial abnormalities are under investigation now.

Another aspect of mitochondrial abnormality in Parkinson's disease relates to the question if free radical damage is involved in nigral cell death or not?

Is free radical damage involved in nigral cell death?

Usually dopamine that is once released into the synaptic space, and re-uptaken into the nerve terminals are the major source of activated oxygen species within nigral cells, as oxidation of dopamine by monoamine oxidase releases hydrogen peroxide, which may in turn produce more toxic hydroxyl radicals by the Fenton reaction according to the following equation (Halliwell, 1989).

$$H_2O_2 + Fe^{2+} \rightarrow {}^{\cdot}OH + OH^- + Fe^{3+}$$

If melanin coexists, melanin reduces Fe^{3+} to Fe^{2+} (Youdim et al., 1989), and this reaction may perpetuate. This is the basic idea of free radical hypothesis regarding the nigral cell death. Another source of activated oxygen species is microsomal NADPH-cytochrome P_{450} reductase (Minakami et al., 1988). As NADPH is oxidized to $NADP^+$ by microsomal cytochrome P_{450} reductase, a pair of electrons are delivered to the oxygen molecules exist within the membrane as well as in the cytoplasm to form superoxide anions. Superoxide anions formed within the membrane will interact with membrane-associated iron to induce lipid peroxidation (Koga and Nakano, 1992). Superoxide anions thus formed within the membrane are more toxic than those liberated into the cytoplasm, oxidizing unsaturated fatty acids composing the intracellular membrane structures. This lipid peroxidation will impair membrane functions, and may contribute to the cell death. In fact increase in the malon dialdehyde was reported in PD (Dexter et al., 1989a). However, report on enzymes regulating the metabolism of activated oxygen species are very controversial (Ambani et al., 1974; Kish et al., 1985; Marttila et al., 1988; Saggu et al., 1989; Johannsen et al., 1991). But decrease in reduced glutathione level in PD supports the presence of oxidative stress (Perry and Yong, 1986). As mentioned in the introduction, iron content is increased in PD, which is consistent with the free radical hypothesis.

Effects of iron on cultured dopaminergic neurons

As discussed in the previous paragraphs, iron accumulation and mito-chondrial complex I deficiency are the two major biochemical abnormalities which appear to have a significance in the pathogenesis of PD. Although these two abnormalities were found independently, they may be mutually related. We have been interested in the possible interactions of iron with mitochondria, and we wanted to see the effects of iron on dopaminergic neurons and their mtDNA. For this purpose, we established primary cultures of fetal rat mesencephalic neurons together with striatal neurons (Mochizuki et al., 1993). The reason why we used co-cultures is that co-cultures simulate the in vivo state better than the mesencephalic culture alone, and are more suitable for the studies to see the effects of toxic substances or to screen the toxic compounds. Briefly cultures were prepared from the ventral mesencephalon and neostriatum of rat embryos at embryonic day 14 as described by Nishi et al. (1990) with modifications. The neostriatum and ventral mesencephalon were dissected in cold calcium- and magnesium-free Hank's balanced salt solution. These tissues were mechanically dissociated and plated into 24-well culture dishes. The cell density in each well was $2.4 \times 10^5/cm^2$. Cultures were maintained in modified Eagle's medium/F12 supplemented with insulin, antibiotics, and 10% fetal bovine serum. For immunohistochemistry cultures grown on cover slips were fixed with 3% paraformaldehyde in PBS, and stained for tyrosine hydroxylase (TH) according to the ABC method. Synthetic dopamine melanin was prepared by auto-oxidation of dopamine according to the method described by Das et al. (1976) with slight modifications. Cultured tissues were incubated with a medium containing 25 μM concentrations of Fe^{3+}-ADP complex with or without synthetic dopamine-melanin for 48 hours. Morphologi-

cal changes of dopaminergic cells were analyzed by immunohistochemistry stained for TH. We also tested the effect of deferoxamine.

Regarding the results, after incubation with iron and synthetic dopamine-melanin for 48 hours, the number of TH-positive cells decreased
markedly without any observable changes in the background population
of TH-negative cells compared to the control culture. Pretreatment with
deferoxamine mesylate prevented this decrease in TH-positive cells.
Then the quantitative evaluation was made by counting the number of
TH-positive cells in an 1-cm^2 area. Fe^{3+} alone decreased the number of
TH-positive cells, however, addition of dopamine-melanin potentiated
this cell loss. Thus we confirmed that iron is toxic to nigral cells in the
concentration that we used, and addition of dopamine-melanin will
enhance the iron-mediated cell loss. We measured the level of thiobarbituric acid (TBA)-reactive products in these cultures. Iron alone did not
increase the amount of TBA-reactive products, however, co-administration of dopamine-melanin significantly increased the TBA-reactive products. Thus toxicity of iron seems to be mediated in part by the lipid
peroxidation with possible involvement of hydroxyl radicals, and in part
by a mechanism other than lipid peroxidation. Details will be published
elsewhere (Mochizuki et al., 1993).

Does iron produce mitochondrial DNA deletion?

Then we wanted to see if iron would induce mtDNA deletion or not.
For this purpose, total DNA was extracted from the mesencephalic-
striatal co-cultures treated with 25 μM of iron for 48 hours, and they
were subjected to the PCR analysis. Rats have circular mtDNA similar to
human mtDNA. We started to see the presence or absence of the deletion
which corresponds to the common 5-kb human deletion encompassing
the ND 5 gene and the ATPase 6/8 gene. In case of human being, this
common deletion has been shown to occur between the 13-base-pair
direct repeats between the ND 5 and ATPase 6/8 genes (Tanaka et al.,
1989). We confirmed the presence of similar direct repeats in rat
mtDNA by a computer-assisted search system. The PCR products were
subjected to agarose-gel electrophoresis and stained with ethidium bro-
mide. In addition to the band derived from the normal mtDNA, two
additional smaller bands were detected. One of them was approximately
5-kb shorter than the normal band in size. These abnormal bands were
barely seen in the PCR products derived from the DNA of control
cultures. Then total DNA extracted from the cultures were digested by a
restriction enzyme, Spe I, which cuts mtDNA at the nucleotide position
10085 between the genes for ND 5 and ATPase 6/8 approximately in the

middle within the expected deletion site between the directly repeated sequences. After the digestion, DNA was subjected to PCR in the same way, and the PCR products were electrophoresed. In this method, normal mtDNA has been digested by the restriction enzyme, and will not be amplified. But deleted mtDNA without the restriction site will be amplified. In fact, the same two shorter bands were observed after the electrophoresis (unpublished observation). Therefore, we intend to study further to answer the question if iron in some way interacts with mtDNA to induce deletions, or not using different techniques.

The mechanism of mitochondrial DNA mutations in Parkinson's disease and ageing: role of iron and free radicals

Recently it has increasingly become apparent that ageing will impose significant damage on mitochondrial DNA which seems to be the basis for mitochondrial respiratory reduction and cell death associated with ageing. Cortopassi et al. (1990) found age-dependent increase in the deleted mtDNA in the myocardium. Similar deletion was also found in diaphragm (Hattori K et al., 1991), the liver (Yen et al., 1991), and the brain (Simonetti et al., 1992; Soong et al., 1992; Corral-Debrinski et al., 1992). Another interesting observation is the age-dependent increase in 8-hydroxyguanosine reported by Hayakawa et al. (1991b). 8-Hydroxy-guanosine is a free radical adduct of guanosine residue of DNA indicating free radical insult to DNA (Cheng et al., 1992). Hydroxyl radicals are implicated as the most likely species responsible for this conversion. Inhibition of mitochondrial respiration by azidothymidine was also found to cause massive conversion of guanosine to 8-hydroxyguanosine (Haya-kawa et al., 1991a). Formation of 8-hydroxydeoxyguanosine will result in G to T and A to C substitutions at the time of DNA duplication (Cheng et al., 1992). Thus oxidative stress within mitochondria will predispose mtDNA to point mutations. Failure of mitochondrial respiration is one of the most potent oxidative stresses, because oxygen molecules cannot completely be reduced to water. As iron content is increased in PD, and iron can probably cross the mitochondrial membranes, there is a good reason to believe that free radical formation is in fact increased within mitochondria of PD patients predisposing mtDNA to point mutations. The deletions of mtDNA found in PD and iron-treated tissue cultures may in part be a consequence of point mutations of mtDNA, be it not a common mutation. These mtDNA mutations may reduce the respiratory efficiency which will in turn enhance the free radical formation, thus forming a vicious cycle. The ultimate result will be energy crisis and neuronal degeneration. The degenerating mitochondria may release iron molecules from the iron-sulfur clusters, which will also contribute to the iron accumulation. In addition, degenerating neurons may not be able to eliminate irons effectively out of the brain. Thus iron increase whether it is primary or secondary will contribute to the progression of the neuronal degeneration.

What is the primary cause of Parkinson's disease?

At present there is no answer to this question. When genetic influence is great, it will manifest as a familial PD. For such families, linkage analysis seems important. For the majority of sporadic cases, perhaps interaction of constitutional factors and environmental factors may play a role in the acquisition of the disease. Among the constitutional factors, poor metabolizers for debrisoquine were found to have two-fold risk of developing PD (Armstrong et al., 1992; Smith et al., 1992). But poor metabolizers do not account for the majority of patients. As an environmental factor, rural living in early life was reported to become a risk factor (Rajuput et al., 1987; Tanner et al., 1989). Another case-control study revealed a certain life style in the early period (less than 40 year-old) characterized by poor intake of fruits, dairy products and alcohol to be associated with a higher risk of obtaining PD (Kondo and Watanabe, 1993).

Uneven distribution of the degenerative changes within substantia nigra is another important issue. The ventro-lateral part of the nigral neurons undergo more severe degeneration (Gibb and Lees, 1991). Neurons in this part project to the mainly putaminal part of the neostriatum which is concerned with motor functions. It can be assumed that neurons concerned with motor control may be metabolically very active. Actually our previous immunohistochemical study for complex I showed that the large oculomotor neurons were the most intensely stained cells among the midbrain structures. The melanin containing nigral neurons were next to the oculomotor neurons in the intensity of immunostaining for electron transfer complexes (Hattori N et al., 1991). Therefore, nigral melanin containing neurons are also considered to be metabolically very active. Metabolically active, particularly oxidatively active neurons owe a higher risk for the exposure to the oxygen derived free radicals. Whatever may be the cause, once the initial insult is given to a subpopulation of nigral neurons, the pathological process may progress automatically thereafter. For instance, dying neurons may evacuate some toxic substances such as lipid peroxides, ferric ions, or neuromelanin. Those substances may be taken up by the adjacent neurons to further extend the area of neuronal damage. Such hypothesis fits well with the very slow and chronic degenerative process of PD. From the therapeutic stand point of view, to interrupt such vicious cycles seems important to retard the progression of the disease, even if such treatment is not a fundamentally curable one. Further studies on this line seem to be important.

Acknowledgements

This study was in part supported by Grant-in Aid-for Priority Areas "Neuronal Death" from Ministry of Culture, Science, and Education. Our studies on mitochondrial DNA analyses on PD patients were performed with the collaboration with Prof. T. Ozawa and Prof. M. Tanaka in Department of Biochemistry of Faculty of Medicine, University of Nagoya. We are very grateful for this collaboration.

References

Albores R, Heafsey EJ, Drucker G, Fields JZ, Collins MA (1990) Mitochondrial respiratory inhibition by N-methylated beta-carboline derivatives structurally resembling N-methyl-4-phenylpyridine. Proc Natl Acad Sci USA 87: 9368–9372

Ambani LM, Van Woert H, Murphy S (1975) Brain peroxidase and catalase in Parkinson disease. Arch Neurol 32: 114–118

Anderson S, Bankier AT, Barrell BG, de Bruijn MHL, Coalson AR, Drouin J, Eperon IC, Nierlich DD, Roe BA, Sanger F, Schreier PH, Smith AJH, Staden R, Young IG (1981) Sequence and organization of the human mitochondrial genome. Nature 90: 457–465

Anderson S, de Bruijn MHL, Coulson AR, Eperon IC, Sanger F, Young IG (1982) Complete sequence of bovine mitochondrial DNA: conserved features of the mammalian mitochondrial genome. J Mol Biol 156: 683–717

Armstrong M, Daly AK, Cholerton S, Bateman DN, Idle JR (1992) Mutant debrisoquine hydroxylation genes in Parkinson's disease. Lancet 339: 1017–1018

Bindoff LA, Birch-Machin M, Cartlidge NEF, Parker WD Jr, Turnbull DM (1989) Mitochondrial function in Parkinson's disease. Lancet ii: 49

Ben-Shachar D, Riederer P, Youdim MBH (1991) Iron-melanin interaction and lipid peroxidation: implications for Parkinson's disease. J Neurochem 57: 1609–1614

Capaldi RA (1982) Arrangement of proteins in the mitochondrial inner membrane. Biochem Biophys Acta 694: 291–306

Caron F, Jacq C, Rouviere-Yaniv (1979) Characterization of a histon-like protein extracted from mitochondria. Proc Natl Acad Sci USA 76: 4265–4269

Cheng KC, Cahill DS, Kasai H, Nishimura S, Loeb LA (1992) 8-Hydroxyguanine, an abundant form of oxidative DNA damage, cause $G \rightarrow T$ and $A \rightarrow C$ substitutions. J Biol Chem 267: 166–172

Chomyn A, Mariottini P, Cleeter MWJ, Ragan CI, Matsuno-Yagi A, Hatefi Y, Doolittle RF, Attardi G (1985) Six unidentified reading frames of human mitochondrial DNA encode components of the respiratory-chain NADH dehydrogenase. Nature 314: 592–597

Chomyn A, Cleeter MWJ, Ragan CI, Riley M, Dolittle RF, Attardi G (1986) URF6, last unidentified reading frame of human mtDNA codes for an NADH dehydrogenase subunit. Science 234: 614–618

Clayton DA (1982) Replication of animal mitochondrial DNA. Cell 28: 693–705

Collins MA, Neafsey EJ, Matsubara K, Cobuzzi RJ Jr, Rollema H (1992) Indol-N-methylated β-carbolinium ions as potential brain-bioactivated neurotoxins. Brain Res 570: 154–160

Corral-Debrinski M, Horton T, Lott MT, Shoffner JM, Beal MF, Wallace DC (1992) Mitochondrial DNA deletions in human brain; regional variability and increase with advanced age. Nature Genet 2: 324–329

Cortopassi GA, Arnheim N (1990) Detection of a specific mitochondrial DNA deletion in tissues of older humans. Nucl Acids Res 18: 6927–6933

Das KC, Abramson MB, Katzman R (1976) Neural pigments: spectroscopic characterization of human brain melanin. J Neurochem 40: 601–605

Dexter DT, Carter CJ, Wells FR, Javoy-Agid F, Agid Y, Lees A, Jenner P, Marsden CD (1989a) Basal lipid peroxidation in substantia nigra is increased in Parkinson's disease. J Neurochem 52: 381–389

Dexter DT, Wells FR, Lees AJ, Agid F, Agid Y, Jenner P, Marsden CD (1989b) Increased nigral iron content and alterations in other metal ions occurring in brain in Parkinson's disease. J Neurochem 52: 1830–1836

Drucker G, Raikoff K, Neafsey EJ, Collins MA (1990) Dopamine uptake inhibitory capacities of beta-carboline and 3,4-dihydro-beta-carboline analogs of N-methyl-4-phenyl-1,2,3,6-tetrahydropyridine (MPTP) oxidation products. Brain Res 509: 125–133

Gibb WRG, Lees AJ (1991) Anatomy, pigmentation, ventral and dorsal subpopulations of the substantia nigra, and differential cell death in Parkinson's disease. J Neurol Neurosurg Psychiatry 54: 388–396

Good PF, Olanow CW, Perl DP (1992) Neuromelanin-containing neurons of the substantia nigra accumulate iron and aluminum in Parkinson's disease: a LAMMA study. Brain Res 593: 343–346

Gross NJ, Getz GS, Rubinowitz M (1969) Apparent turnover of mitochondrial deoxyribonucleic acid and mitochondrial phospholipids in the tissue of rat. J Biol Chem 244: 1552–1562

Halliwell B (1989) Oxidants and the central nervous system: some fundamental questions. Acta Neurol Scand 126: 23–33

Hatefi Y (1985) The mitochondrial electron transport and oxidative phosphorylation system. Annu Rev Biochem 54: 1015–1069

Hattori K, Tanaka M, Sugiyama S, Obayashi T, Ito T, Satake T, Hanaki Y, Asai J, Nagano M, Ozawa T (1991) Age-dependent increase in deleted mitochondrial DNA in the human heart: possible contributory factor to presbycardia. Am Heart J 121: 1735–1742

Hattori N, Tanaka M, Ozawa T, Mizuno Y (1991) Immunohistochemical studies on Complex I, II, III, and IV of mitochondria in Parkinson's disease. Ann Neurol 30: 563–571

Hayakawa M, Ogawa T, Sugiyama S, Tanaka M, Ozawa T (1991a) Massive conversion of guanosine to 8-hydroxy-guanosine in mouse liver mitochondrial DNA by administration of azidothymidine. Biochem Biophys Res Commun 176: 87–93

Hayakawa M, Torii K, Sugiyama S, Tanaka M, Ozawa T (1991b) Age-associated accumulation of 8-hydroxydeoxyguanosine in mitochondrial DNA of human diaphragm. Biochem Biophys Res Commun 179: 1023–1029

Heikkila RE, Nicklas WL, Vyas I, Duvoisin RC (1985) Dopaminergic toxicity of rotenone and the 1-methyl-4-phenylpyridinium ion after their stereotaxic administration to rats: implication for the mechanism of 1-methyl-4-phenyl-1,2,5,6-tetrahydropyridine toxicity. Neurosci Lett 62: 389–394

Hirsch EC, Brandel JP, Galle P, Javoy-Agid F, Agid Y (1991) Iron and aluminum increase in the substantia nigra of patients with Parkinson's disease: an X-ray microanalysis. J Neurochem 56: 446–451

Ikebe S, Tanaka M, Ohno K, Sato W, Hattori K, Kondo T, Mizuno Y, Ozawa T (1990) Increase of deleted mitochondrial DNA in the striatum in Parkinson's disease and senescence. Biochem Biophys Res Commun 170: 1044–1048

Jellinger K, Paulus W, Grundke-Iqbal P, Riederer P, Youdim MBH (1990) Brain iron and ferritin in Parkinson's and Alzheimer's disease. J Neural Transm [P-D Sect] 2: 327–340

Jellinger K, Kienzl E, Rumpelmair G, Riederer P, Stachelberger H, Ben-Shachar D, Youdim MBH (1992) Iron-melanin complex in substantia nigra parkinsonian brains: an X-ray microanalysis. J Neurochem 59: 1168–1171

Johannsen P, Velander G, Mai J, Thorling EB, Dupont E (1991) Glutathione peroxidase in early and advanced Parkinson's disease. J Neurol Neurosurg Psychiatry 54: 679–682

Kagawa Y, Sone N (1979) DCCD-sensitive ATPase (TF$_0$ · F$_1$) from a thermophilic bacterium: purification, dissociation into functional subunits, and reconstitution into vesicles capable of energy transformation. In: Fleischer S, Packer L (eds) Methods in enzymology, vol LV. Academic Press, New York, pp 364–372

Kish SJ, Morito C, Hornykiewicz O (1985) Glutathione peroxidase activity in Parkinson's disease. Neurosci Lett 58: 343–346

Koga S, Nakano M (1992) A high involvement of O$_2$-possibly generated in inner membranes for iron-induced microsomal lipid peroxidation. Biochem Biophys Res Commun 186: 1087–1093

Kondo K, Watanabe K (1993) Lifestyles, risk factors, and inherited predispositions in Parkinson's disease: preliminary report of a case-control study. In: Narabayashi H, Nagatsu T, Yanagisawa N, Mizuno Y (eds) Advances in neurology, vol 60. Raven Press, New York, pp 346–351

Krige D, Carroll MT, Cooper JM, Marsden CD, Schapira AHV (1992) Platelet mitochondrial function in Parkinson's disease. Ann Neurol 32: 782–788

Kunkel TA, Loeb LA (1981) Fidelity of mammalian DNA polymerase. Science 213: 765–767

Lestienne P, Nelson J, Riederer P, Jellinger K, Reichmann H (1990) Normal mitochondrial genome in brain from patients with Parkinson's disease and complex I defect. J Neurochem 55: 1810–1812

Linnane AW, Marzuki S, Ozawa T, Tanaka M (1989) Mitochondrial DNA mutations as an important contributor to ageing and degenerative diseases. Lancet i: 642–645

Linnane AW, Baumer A, Maxwell RJ, Preston H, Zhang C, Marzuki S (1990) Mitochondrial gene mutation: the ageing process and degenerative diseases. Biochem Int 22: 1067–1076

Mann VM, Cooper JM, Krige D, Daniel SE, Schapira AH, Marsden CD (1992) Brain skeletal muscle and platelet homogenate mitochondrial function in Parkinson's disease. Brain 115: 333–342

Marttila RJ, Lorentz H, Rinne UK (1988) Oxygen toxicity protecting enzymes in Parkinson's disease: increase of superoxide dismutase-like activity in the substantia nigra and basal nucleus. J Neurol Sci 86: 321–331

Merle P, Kadenbach B (1980) The subunit composition of mammalian cytochrome c oxidase. Eur J Biochem 105: 499–507

Minakami H, Arai H, Nakano M, Sugioka K, Suzuki S, Sotomatsu A (1988) A new and suitable reconstructed system for NADPH-dependent microsomal lipid peroxidation. Biochem Biophys Res Commun 153: 973–978

Mizuno Y, Sone N, Saitoh T (1986) Dopaminergic neurotoxin, MPTP and MPP+, inhibit mitochondrial NADH-ubiquinone oxidoreductase activity. Proc Jpn Acad Ser B 62: 261–263

Mizuno Y, Sone N, Saitoh T (1987) Effects of 1-methyl-4-phenyl-1,2,3,6-tetrahydropyridine and 1-methyl-4-phenylpyridinium ion on activities of enzymes in the electron transport system in mouse brain. J Neurochem 48: 1787–1793

Mizuno Y, Ohta S, Tanaka M, Takamiya S, Suzuki K, Sato T, Oya H, Ozawa T, Kagawa Y (1989) Deficiencies in complex I subunits of the respiratory chain in Parkinson's disease. Biochem Biophys Res Commun 163: 1450–1455

Mizuno Y, Suzuki K, Ohta S (1990) Postmortem changes in mitochondrial respiratory enzymes in brain and a preliminary observation in Parkinson's disease. J Neurol Sci 96: 49–57

Mochizuki H, Nishi K, Mizuno Y (1993) Iron-melanin complex is toxic to dopaminergic neurons in a nigrostriatal co-culture. Neurodegeneration 2: 1–7

Nagatsu T, Yoshida M (1988) An endogenous substance of the brain, tetrahydroisoquinoline, produces parkinsonism in primates with decreased dopamine, tyrosine hydroxylase and biopterine in the nigrostriatal regions. Neurosci Lett 87: 178–182

Nakagawa-Hattori Y, Yoshino H, Kondo T, Mizuno Y, Horai S (1992) Is Parkinson's disease a mitochondrial disorder? J Neurol Sci 107: 29–33

Naoi M, Matsuura S, Takahashi T, Nagatsu T (1989a) A N-methyltransferase in human brain catalyzes N-methylation of 1,2,3,4-tetrahydroisoquinoline into N-methyl-1,2,3,4-tetrahyroisoquinoline, a precursor of a dopaminergic neurotoxin N-methylisoquinolinium ion. Biochem Biophys Res Commun 161: 1213–1219

Naoi M, Matsuura S, Parvez H, Takahashi T, Hirata Y, Minami M, Nagatsu T (1989b) Oxidation of N-methyl- 1,2,3,4-tetrahydroisoquinoline into the N-methyl-isoquinolinium ion by monoamine oxidase. J Neurochem 52: 653–655

Nicklas WL, Vyas I, Heikkila RE (1985) Inhibition of NADH-linked oxidation in brain mitochondria by 1-methyl-4-phenyl-pyridine, a metabolite of the neurotoxin, 1-methyl-4-phenyl-1,2,5,6-tetrahydropyridine. Life Sci 36: 2503–2508

Nishi K, Akins PT, Surmier DJ, Kitai ST (1990) Muscarinic regulation of cyclic AMP metabolism in rat neostriatal cultures. Brain Res 543: 111–116

Niwa T, Takeda N, Kaneda N, Hashizume Y, Nagatsu T (1987) Presence of tetrahydroisoquinoline and 2-methyl-tetrahydroisoquinoline in parkinsonian and normal human brains. Biochem Biophys Res Commun 144: 1084–1089

Ohta S, Kohno M, Makino Y, Tachikawa O, Hirobe M (1987) Tetrahydroisoquinoline and 1-methyl-tetrahydroisoquinoline are present in the human brain: relation to Parkinson's disease. Biomed Res 8: 453–456

Ozawa T, Tanaka M, Ikebe S, Ohno K, Kondo T, Mizuno Y (1990) Quantitative determination of deleted mitochondrial DNA relative to normal DNA in parkinsonian striatum by a kinetic PCR analysis. Biochem Biophys Res Commun 172: 483–489

Parker WD Jr, Boyson SJ, Parks JK (1989) Abnormalities of the electron transport chain in idiopathic Parkinson's disease. Ann Neurol 26: 719–723

Perry TL, Yong VW (1986) Idiopathic Parkinson's disease, progressive supranuclear palsy and glutathione metabolism in the substantia nigra of patients. Neurosci Lett 67: 269–274

Ragan CI (1987) Structure of NADH-ubiquinone reductase (Complex I). In: Lee CP (ed) Current topics in bioenergetics: structure, biogenesis, and assembly of energy transducing enzyme systems. Academic Press, San Diego, pp 1–36

Ragan CI, Galante YM, Hatefi Y (1982) Purification of three iron-sulfur proteins from the iron-protein fragment of mitochondrial NADH-ubiquinone oxidoreductase. Biochemistry 21: 2518–2514

Rajput AH, Uitti RJ, Stern W, Laverty K, O'Donnell K, O'Donnel D, Yuen WK, Dua A (1987) Geography, drinking water chemistry, pesticides and herbicides and the etiology of Parkinson's disease. Can J Neurol Sci 14: 414–418

Ramsay RR, Salach JI, Dadgar J, Singer TP (1986) Inhibition of mitochondrial NADH dehydrogenase by pyridine derivatives and its possible relation to experimental and idiopathic parkinsonism. Biochem Biophys Res Commun 135: 269–275

Reichmann H, Riederer P, Seufert S (1990) Disturbances of the respiratory chain in brain from patients with Parkinson's disease. Mov Disord 5: 28 (abstr)

Reichmann H, Lestienne P, Jellinger K, Riederer P (1993) Parkinson's disease and the electron transport chain in postmortem brain. In: Narabayashi H, Nagatsu T, Yanagisawa N, Mizuno Y (eds) Advances in neurology, vol 60. Raven Press, New York, pp 297–299

Richter C, Park J-W, Ames BN (1988) Normal oxidative damage to mitochondrial and nuclear DNA is extensive. Proc Natl Acad Sci USA 85: 6465–6467

Riederer P, Sofic E, Rausch W-D, Schmidt B, Reynolds GP, Jellinger K, Youdim MBH (1989) Transition metals, ferritin, glutathione, and ascorbic acid in parkinsonian brains. J Neurochem 52: 515–520

Saggue H, Cooksey J, Dexter D, Wells FR, Lees A, Jenner P, Marsden CD (1989) A selective increase in particulate superoxide dismutase activity in parkinsonian substantia nigra. J Neurochem 53: 692–697

Schapira AHV, Cooper JM, Dexter D, Jenner P, Clark JB, Marsden CD (1989) Mitochondrial complex I deficiency in Parkinson's disease. Lancet i: 1269

Schapira AHV, Cooper JM, Dexter D, Clark JB, Jenner P, Marsden CD (1990a) Mitochondrial complex I deficiency in Parkinson's disease. J Neurochem 54: 823–827

Schapira AHV, Holt IJ, Sweeney M, Harding AE, Jenner P, Marsden CD (1990b) Mitochondrial DNA analysis in Parkinson's disease. Mov Disord 5: 294–297

Shoffner JM, Watts RL, Juncos JL, Torroni A, Wallace DC (1991) Mitochondrial oxidative phosphorylation defects in Parkinson's disease. Ann Neurol 30: 332–339

Simonetti S, Chen X, DiMauro S, Schon EA (1992) Accumulation of deletions in human mitochondrial DNA during normal aging: analysis by quantitative PCR. Biochim Biophys Acta 1180: 113–122

Singer TP, Ramsay RR (1990) Mechanism of the neurotoxicity of MPTP. FEBS Lett 274: 1–8

Smith CA, Gough AC, Leigh PN, Summers BA, Harding AE, Maranganore DM, Sturman SG, Schapira AV, Williams AC, Spurr NK, Wolf CR (1992) Debrisoquine hydroxylase gene polymorphism and susceptibility to Parkinson's disease. Lancet 339: 1375–1377

Sofic E, Paulus W, Jellinger K, Riederer P, Youdim MBH (1991) Selective increase of iron in substantia nigra zona compacta of parkinsonian brains. J Neurochem 56: 978–982

Soong NW, Hinton DR, Cortopassi G, Amheim N (1992) Mosaicism for a specific somatic mitochondrial DNA mutation in adult human brain. Nature Genet 2: 318–323

Suzuki K, Mizuno Y, Yamauchi Y, Nagatsu T, Yoshida M (1992) Selective inhibition of complex I by N-methylisoquinolinium ion and N-methyl-1,2,3,4-tetrahydroisoquinoline in isolated mitochondria prepared from mouse brain. J Neurol Sci 109: 219–223

Tanaka M, Sato W, Ohno K, Yamamoto T, Ozawa T (1989) Direct sequencing of deleted mitochondrial DNA in myopathic patients. Biochem Biophys Res Commun 184: 156–163

Tanner CM, Chen B, Wang W, Peng M, Liu Z, Liang X, Kao LC, Gilley DW, Goeta CG, Schoenberg BS (1989) Environmental factors and Parkinson's disease: a case-control study in China. Neurology 39: 660–664

Tasaki Y, Makino Y, Ohta S, Hirobe M (1991) 1-Methyl-1,2,3,4-tetrahydroisoquinoline, decreasing in 1-methyl-4-phenyl-1,2,3,6-terahydropyridine-treated mouse, prevents parkinsonism-like behavior abnormalities. J Neurochem 57: 1940–1943

Walker JE (1992) The NADH: ubiquinone oxidoreductase of respiratory chains. Quart Rev Biophys 25: 253–324

Walker JE, Arimendi JM, Dupuis A, Fearnley IM, Finel M, Medd SM, Pilkington SJ, Runswick MJ, Skehel JM (1992) Sequences of 20 subunits of NADH: ubiquinone oxidoreductase from bovine heart mitochondria: application of a novel strategy for sequencing proteins using the polymerase chain reaction. J Mol Biol 226: 1051–1072

Weiss H, Friedrich T (1991) Redox linked proton translocation by NADH-ubiquinone reductase (complex I). J Bioenerget Biomembr 23: 743–754

Yen T-C, Su J-H, King K-L, Wei Y-H (1991) Ageing-associated 5 kb deletion in human liver mitochondrial DNA. Biochem Biophys Res Commun 178: 124–131

Yoshida M, Niwa T, Nagatsu T (1990) Parkinsonism in monkeys produced by chronic administration of an endogenous substance of the brain, tetrahydro-isoquinoline: the behavioral and biochemical changes. Neurosci Lett 119: 109–113

Yoshino H, Nakagawa-Hattori Y, Kondo T, Mizuno Y (1992) Mitochondrial complex I and complex II activities of lymphocytes and platelets in Parkinson's disease. J Neural Transm [P-D Sect] 4: 27–34

Youdim MBH, Ben-Shachar D, Riederer P (1989) Is Parkinson's disease a progressive siderosis of substantia nigra resulting in ion and melanin induced neurodegeneration? Acta Neurol Scand 126: 47–54

Correspondence: Dr. Y. Mizuno, Department of Neurology, Juntendo University School of Medicine, 2-1-1 Hongo, Bunkyo, Tokyo 113, Japan.

Consequences of intrastriatally administrated FeCl₃ and 6-OHDA without and after transient cerebral oligemia on behaviour and navigation

A free radical related tissue damage?

C. Heim[2], W. Kolasiewicz[1,a], T. Jaros[1,a], and K.-H. Sontag[1]

[1] Department of Pharmacology, Max-Planck-Institute for Experimental Medicine, Göttingen, and Veterinär-biologisches Labor, Northeim, Federal Republic of Germany
[2] Department of Psychiatry, University of Göttingen, Göttingen, Federal Republic of Germany

Summary

6-OHDA (6 or 8 μg) unilaterally applied into the ventrolateral striatum induces neurotoxic effects which lead to unilateral hyperactivity triggering contralateral turning and rotations after apomorphine administration. Treatment with alpha-tocopherol prevents the hypersensitive reaction. The sensitivity of the cerebral tissue to apomorphine following 6-OHDA treatment is enhanced when 7.5 μg FeCl₃ is coadministered with 6-OHDA. Only when the combination of 6-OHDA and FeCl₃ is administered the escape latency to find a hidden platform in a water maze-test increased as measured 12 weeks later. The night activity of such treated animals was markedly reduced. Remarkable effects of unilaterally applied FeCl₃ (0.06–1.5 ug) were observed in old rats, and in rats after transient cerebral oligemia. Apomorphine treatment to such animals induced rotations. It is possible, therefore, that mild to moderate transient or permanent local oxygen deficits together with iron cause progressive damage to vulnerable cerebral tissue. Such an effect would be comparable to the neurotoxicity of 6-OHDA possibly involve free radicals and lipid peroxidation.

Introduction

Nigro-striatal DAergic neurons of rats are susceptible to the presence of ionic iron (Ben Shachar and Youdim, 1991). Iron ions are free radicals;

---

[a] On leave: Institute of Pharmacology, Polish Academy of Science, Krakow, Poland

iron (II) ions (Fe^{2+}) can take part in electron transfer reactions with molecular oxygen (Halliwell and Gutteridge, 1989). Iron plays an important role in lipid peroxidation. Pure lipid peroxides are fairly stable at physiological temperatures, but in the presence of transition metal complexes their decomposition is greatly accelerated. The overall effects of lipid peroxidation are to decrease membrane fluidity, increase the "leakiness" of the membrane to substances that do not normally cross it (such as Ca^{2+} ions) and inactivate membrane bound enzymes. Added iron ion "stimulates" further lipid peroxidation (Halliwell and Gutteridge, 1989).

Lipid peroxidation does appear to play a major role in arterio-sclerosis and in brain and spinal cord during ischemic processes and reperfusion (Siesjö et al., 1992).

The neurotoxic effect of 6-OHDA is thought to involve the production of free radicals (Cohen and Heikkila, 1974) which induces behavioural abnormalities like contralateral rotation after subcutaneously apomorphine treatment. 6-OHDA injected into the brain causes specific damage of catecholamine nerve terminals (Jonsson, 1980). This is due to the specific uptake of 6-OHDA by catecholaminergic neurones and the subsequent formation of O_2^-, hydrogen peroxide and OH^-.

Pretreatment of animals with alpha-tocopherol causes significant attenuation of the toxicologically 6-OHDA induced abnormal behaviour. Alpha-tocopherol donates labil hydrogen to membranes so terminating chain reactions of peroxidation caused by scavenging chain propagating radicals (Halliwell and Gutteridge, 1989); this is how antioxidants may function beneficially in the treatment of neurodegenerative disorders.

Several studies suggest a causal relationship between the pathogenesis of ischemic brain damage and oxygen derived free radicals during reperfusion or after permanent ischemia (Flamm et al., 1978; Halliwell, 1978; Siesjö, 1981; Yoshida et al., 1982; Halliwell and Gutteridge, 1985; Kogure et al., 1985; Kirsch et al., 1987; Watson and Ginsberg, 1989; Haba et al., 1991; Boisvert and Schreiber, 1992). Many studies have shown free radical scavengers and antagonists of exitatory amino acids to be effective anti-ischemia agents (Yamamoto et al., 1983; Asano et al., 1984; Ito et al., 1986; Abe et al., 1988; Young et al., 1988; Hall and Yonkers, 1988; Kurumaji et al., 1989; Liu et al., 1989; Martz et al., 1989; Hall, 1990; Meldrum, 1990; Betz, 1990; Bullock et al., 1990; Sheardown et al., 1990; Diemer et al., 1992; Ginsberg et al., 1992; Lekieffre et al., 1992; Lippert et al., 1992; Meldrum et al., 1992; Moncada et al., 1992; Oh and Betz, 1992; Wahl et al., 1992).

But also the transient reduction of cerebral blood flow to oligemic levels by bilateral clamping of carotid arteries (BCCA) for 60 min in normotensive Wistar rats is known to lead to the stimulation of the glutamatergic system of the striatum as well as an increase in lipid peroxidation in the frontal cortex and striatum determined 2 weeks after the insult (Läer et al., 1992, 1993; Melzacka et al., 1992; Sontag et al., 1992).

It is possible, therefore, that the effects of intrastriatally administered iron, or iron together with 6-OHDA, or iron during oxygen deprivation act by a common mechanism, i.e. production of reactive oxygen species (free radical production) which initiates or increases lipid peroxidation. The outcome of this may be detectable behavioural deficiencies.

Material and methods

Surgery

Male wistar rats aged 3–4, 14–16 and 26–28 months housed under controlled temperature and 12/12 light/dark cycle had free access to food and water. Rats were anaesthetized with 60 mg/kg pentobarbital (Nembutal, Sanofi, France) for surgery. Different groups were used: 1. naive group of animals received no surgery; 2. sham operated group of rats were intrastriatally injected with 2 µl 0.02% ascorbic acid plus 2 0.9% NaCl; 3. groups of rats were intrastriatally injected with 6 µg 6-OHDA in 2 µl 0.02% ascorbic acid or 2 µl 0.02% ascorbic acid alone for control (sham); subgroups received either alpha-tocopherol 100 mg/kg body weight or the same volume of 0.9% NaCl (0.5 ml/100g body weight) intraperitoneally for 4 weeks prior to surgery. 4. Groups of animals were intrastriatally injected with 8 µg 6-OHDA in 2 µl N₂-bubbled NaCl or 2 µl

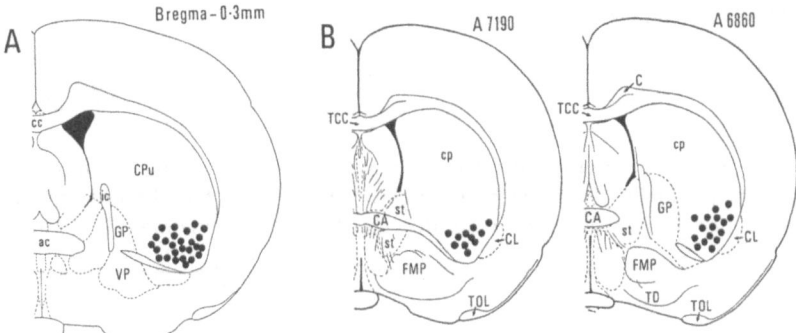

Fig. 1. Schematic drawing showing the injection sites; coordinates according to Paxinos and Watson (1982), bregma -0.3 (**A**); coordinates according to König and Klippel (1963), A 7190 and A 6860 (**B**); *cp* (striatum) resp. caudate-putamen; *GP* globus pallidus; *VP* ventral pallidum; *CL* Claustrum; *CC* truncus corporis callosi TCC; *FMP* fasciculus medialis proencephali; *TOL* tractus olfactorius lateralis; *ac* or *CA* resp. commissura anterior; *ic* capsula interna; *st* nucleus interstitialis striae terminalis

0.02% ascorbic acid; subgroups received either an additional intrastriatal injection of 7.5 µg $FeCl_3$ in 2 µl NaCl or 2 µl NaCl alone. 5. Groups of animals after bilateral clamping of the carotid arteries (see below) and intrastriatally injected with 0.06, 0.3 or 1.5 µg $FeCl_3$ dissolved in 2 µl saline with buffer (pH 7.4) (see Sengstock et al. 1992), or 2 µl solvent for control, 1 week later. Injections were performed unilaterally into the ventro-lateral striatum using coordinates according to Paxinos and Watson (1982) (Fig. 1).

A group of aged rats (14–16 and 26–28 months) received small amounts of $FeCl_3$ (1.5, 0.3 and 0.06 µg) alone or in combination with a transient cerebral oligemia by occlusion of both carotid arteries. These rats of different age (see results) were subjected to occlusion of the common carotid arteries for 60 min in pentobarbital anaesthesia (Nembutal, 60 mg/kg i.p.). The mean arterial blood pressure was normal during occlusion and reperfusion and the local cerebral blood flow within the striatum was not critical and did not fall to an ischemic level (Block et al., 1993). 14 days after this procedure a significant elevation of ascorbate-induced malondialdehyd formation could be measured (Melzacka et al. 1992; Sontag et al., 1992). Sham control animals had only their vessels prepared but not clamped. Rectal temperature was 37°C. Control animals (naive) received neither anaesthesia nor surgery.

Drugs

0.4 mg or 2 mg/kg apomorphine (Sigma, St. Louis, Missouri, USA) were subcutaneously injected to test spontaneous and rotation behaviour. Animals treated with 6-OHDA (RBI, Natick, Massachusetts, USA) received pargyline (50 mg/kg; Sigma, St. Louis, Missouri, USA) i.p. and desmethyl-imipramine (25 mg/kg; Sigma, St. Louis, Missouri, USA) i.p. Alpha-tocopherolacetat (E-Vicotrat®, Heyl, Berlin, Germany) was injected i.p. daily for 4 weeks prior to intrastriatal applications.

Behavioural tests

Spontaneous activity and rotation behaviour was tested using an autotrack-system (Columbus Instruments, Columbus, Ohio, USA), but the rotating was controlled by the experimentator directly. Only total circlings (360°) were counted.

Swimming pool

The rats were tested in a black circular pool (diameter 100 cm) filled with 26 °C water. A perspex transparent platform (diameter 6 or 8 cm) was mounted 1 cm below the water level and was therefore not visible to the rats during swimming.

Learning-set task

Following the description by Whishaw (1987) four platform locations were used and the platform was moved each day to one of these locations according to a designated sequence (see Fig. 6). As described by Whishaw, the platform positions were chosen to frustrate a number of none-place learning strategies that rats may adopt. A rat may attempt to locate the platform by swimming in a circular path around the pool: if this strategy is adopted a platform located at position 3, in the centre will not be found. A rat may turn away from the wall and swim at a given angle: this strategy will not help to reach platform location 4, which is immediately adjacent to a start position and located slightly closer to the wall than position 1 and 2, and which requires that the rats swim towards the centre of the pool to locate it. A rat may concentrate on swimming in a quadrant or half of the pool: the asymmetric locations of the platforms will limit the utility of this strategy. Four starting locations were used: north-east, south-east, south-west and north-west. The rats were gently placed into the water facing and touching the wall of the pool at these starting points.

Testing was conducted on consecutive days. Each rat received 16 trials on each day. If on a particular trial a rat found the platform, it was permitted to remain there for 10 s. A trial was terminated after 120 s if a rat failed to find the platform. Trials were given in pairs ("twin"-trials). The second trial of each pair was given immediately after the 10 s stay on the platform, and the same starting location was used. At the end of the second of each pair of trials, the rat was returned to its home cage and some minutes elapsed before the next pair of trials was begun from a new starting location, ect. Trial pairs ("twin"-trials) were used so that the rats started from each of the four locations on each of the first 4 pairs of trials (trial blocks) and each of the second pairs of trials. The sequence in which starting positions were used was random (see Whishaw, 1987). Videorecordings were made continuously during the trials for further evaluations.

Results

The effect of 6 μg of 6-OHDA injected unilaterally into the ventro-lateral striatum was tested one week after surgery during the exploration phase (Fig. 2) and 3 and 13 weeks after surgery with apomorphine (Fig. 3). 6-OHDA decreased the spontaneous activity significantly (Fig. 2). As shown in Fig. 3 A 0.4 mg/kg apomorphine induced typical contralateral rotations. 13 weeks after surgery an enhanced rotating behaviour was still evident when the same animals received 2 mg/kg apomorphine, demonstrating the longlasting – possibly chronic – defect of the striato-nigral DAergic system.

The registration of the locomotor activity 9–10 weeks after surgery showed a significant decrease in the distance travelled (DT) (Fig. 4) and

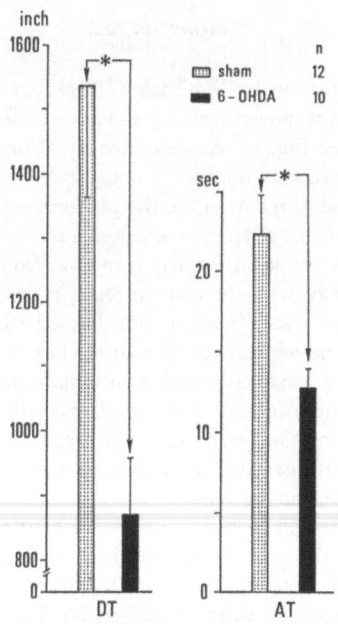

Fig. 2. Spontaneous behaviour (locomotor activity) during the exploration phase 1 week after an intrastriatal injection of 6 μg 6-OHDA dissolved in 2 μl 0.02% ascorbic acid or 2 μl 0.02% ascorbic acid alone (sham). Registration for 30 min between 9 a.m. and 1 p.m. *DT* distance travelled; *AT* time of ambulatory movements. Statistics: One-way-ANOVA and Newman-Keuls-test; means ± SEM

ambulatory time (AT) (not shown) in the group of rats treated with 8 μg of 6-OHDA and 7.5 mg/kg $FeCl_3$ (Fig. 4).

The effect of intrastriatal iron was much more remarkable in the following experiment 18 and 26 weeks after surgery. Significantly more rotations occured for animals treated with 6-OHDA and $FeCl_3$ than with 6-OHDA alone. 7.5 μg $FeCl_3$ was also more effective than just 8 μg 6-OHDA (Fig. 5).

The cognitive performance of 6-OHDA lesioned animals with and without coadministration of $FeCl_3$ and tested 12 weeks after surgery in the water maze is demonstrated in Fig. 6.

The time to reach the hidden platform (escape latency) in south position (S) during 2 trial blocks (= 16 trials) at day 1 was much higher for the group of animals treated with a combination of $FeCl_3$ and 6-OHDA than for groups receiving only 6-OHDA (Fig. 6). The increase in escape latency caused by an additional injection of $FeCl_3$ was persistently present during all trial blocks (Fig. 6).

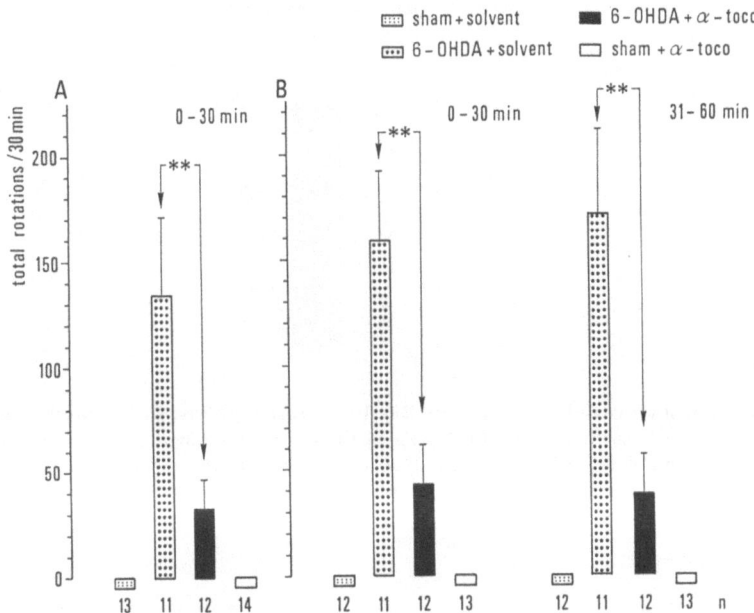

Fig. 3. Registration of contralateral rotations **A** 3 weeks after surgery following subcutaneous injection of 0.4 mg/kg apomorphine s.c. and **B** 13 weeks after surgery following 2 mg/kg apomorphine s.c. *Sham + solvent* intrastriatal injection of 2 µl 0.02% ascorbic acid following a pretreatment with 0.5 ml/100 g body weight 0.9% NaCl; *6-OHDA + solvent* intrastriatal injection of 6 µg 6-OHDA in 2 µl 0.02% ascorbic acid following a pretreatment with 0.5 ml/100g body weight i.p. NaCl; *6-OHDA + α-toco* intrastriatal injection of 6 µg 6-OHDA in 2 µl 0.02% ascorbic acid following pretreatment with α-tocopherol; *sham + α-toco* intrastriatal injection of 2 µl 0.02% ascorbic acid following pretreatment with α-tocopherol. α-tocopherol (100 mg/kg) or NaCl (0.5 ml/100 g body weight) were administered i.p. for 4 weeks daily prior to surgery. Time of observation in **A** 30 min and in **B** 30–60 min after apomorphine injection. Circlings of 360° as means ± SEM. Statistics: ** p < 0.01; one-way-ANOVA and Newman-Keuls-test; n = number of animals used

The sensitivity to intrastriatally injected FeCl₃ increased dramatically in old rats (26–28 month old) and in adult animals (14–16 month old) after a 60 min transient reduction of cerebral blood flow produced by occlusion of both carotid arteries in pentobarbital anaesthesia. For 26–28 month old rats without transient cerebral blood flow reduction, 7.5 µg of FeCl₃ was toxic (6 out of 8 rats operated died after spontaneous circling, akinetic behaviour and hunch back posture). Animals injected with 1.5 µg FeCl₃ intrastriatally one week after 60 min of cerebral oligemia and a treatment with 0.4 mg/kg apomorphine 1 week later leads to contralateral rotations (194 rotations in 90 min). This long-lasting defect could be demonstrated 3 weeks after iron administration with 87 contra-

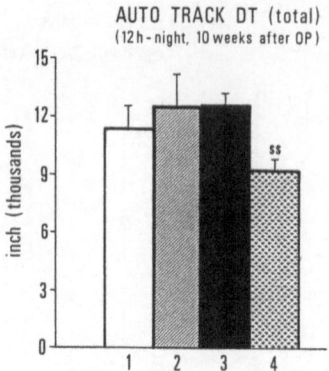

Fig. 4. Night activity (distance travelled, DT) 10 weeks after surgery. *1* naive rats without surgery; *2* 0.02% ascorbic acid (2 µl) plus 2 µl NaCl; *3* 6-OHDA (8 µg in 2 µl ascorbic acid) plus 2 µl NaCl; *4* 6-OHDA (8 µg in 2 µl ascorbic acid) plus FeCl$_3$ 7.5 µg in 2 µl NaCl 2–4 were injected unilaterally into rats' ventrolateral striatum. Statistics: \$\$ p < 0.02 vs. 6-OHDA + NaCl; one-way-ANOVA and Newman-Keuls-test; means ± SEM, n = number of animals used

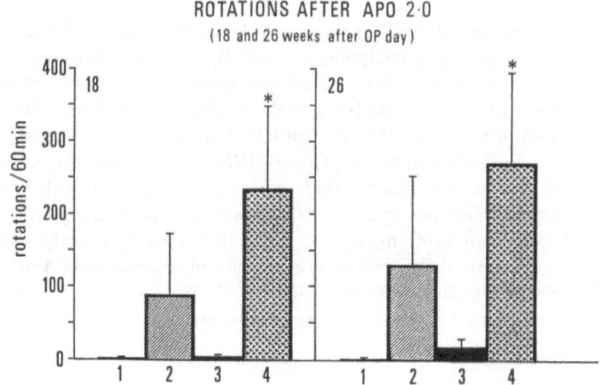

Fig. 5. Apomorphine induced rotations (2 µg/kg s.c.) 18 or 26 weeks after surgery. *1* 0.02% ascorbic acid (2 µl) plus 2 µl NaCl; *2* 2 µl 0.02% ascorbic acid plus 7.5 µg FeCl$_3$ in 2 µl NaCl; *3* 6-OHDA (8 µg in 2 µl ascorbic acid) plus 2 µl NaCl; *4* 6-OHDA (8 µg in 2 µl ascorbic acid) plus FeCl$_3$ 7.5 µg in 2 µl NaCl. 1–4 were injected unilaterally into rats' ventrolateral striatum. Statistics: * p < 0.05 vs 6-OHDA + NaCl; one-way-ANOVA and Newman-Keuls-test; means ± SEM, n = number of animals used

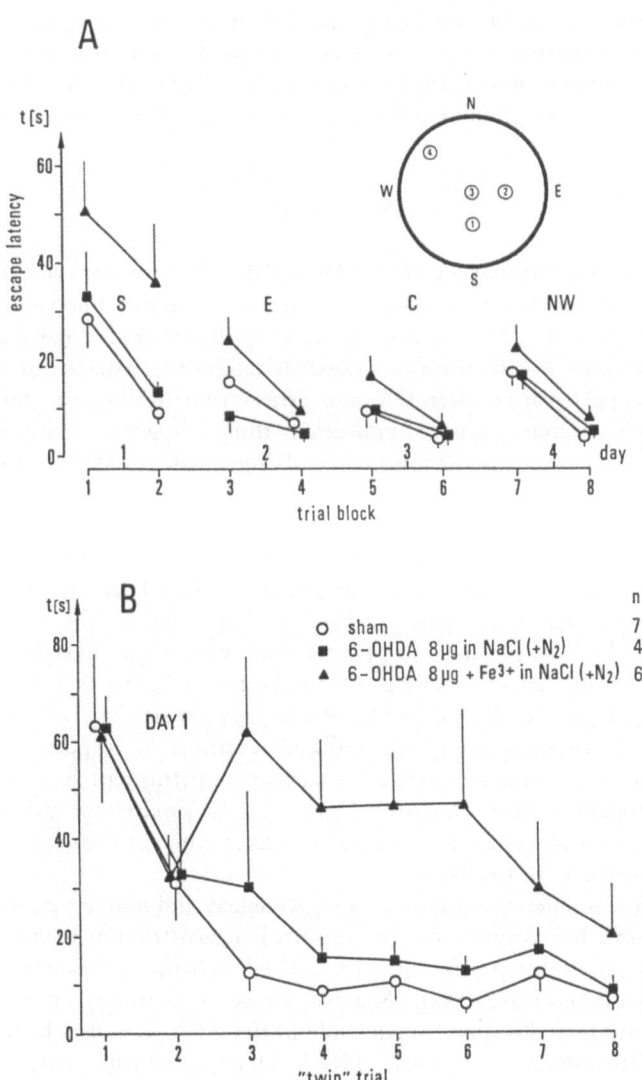

Fig. 6. A Time in [s] to reach the platform (escape latency) during 64 trials within 4 days 12 weeks after surgery. One trial block includes 8 trials, i.e. 2 trials ("twin"-trials) starting from each of the 4 points of the compass (SO, NO, SO, SW), resp., semi-randomly varied. Sequence of platform positions: south [S]: day1; east [E]: day 2; centre [C]: day 3; northwest [NW]: day 4. **B** Escape latencies on day 1 (platform-position south [S]) during 16 trials, i.e. 2 × 2 pairs of trials ("twin"-trials) from each of the 4 starting positions (SO, NO, NW, SW), resp., semi-randomly varied. Statistics: Significant with p < 0.05: 6-OHDA + Fe in NaCl (+ N₂) vs. sham and vs. 6-OHDA in NaCl (+ N₂), ANOVA plus Gabriel's multiple comparison of means; values are means ± SEM; n = number of animals used

lateral rotations in 90 min being induced by 0.4 mg/kg apomorphine s.C. Three of seven rats which were injected with 0.3 µg FeCl₃ one week after the oligemic event turned to circles when they were tested with the mentioned amount of apomorphine one week after FeCl₃ administration.

Discussion

Application of 6 or 8 µg 6-OHDA unilaterally into the ventrolateral striatum induces neurotoxic effects within the DAergic innervated ventrolateral striatum. This was demonstrated by the pharmacological action of apomorphine which stimulates DAergic receptors. Postsynaptic receptors are supersensitive after DAergic denervation leading to unilateral hyperactivity; apomorphine application thus triggering controlateral turning and rotations of the affected rats (Ungerstedt, 1971). The finding that alpha-tocopherol pretreatment counteracts the hypersensitive reaction to apomorphine suggests that the alpha-tocopherol content of the tissue was sufficient to prevent the production of damaging reactive oxygen species (Halliwell and Gutteridge, 1989). Furthermore, it is shown that intrastriatally injected FeCl₃ (7.5 µg) elevates the sensitivity of the cerebral tissue more to apomorphine than 8 µg 6-OHDA. The most remarkable effect of iron could be observed in old rats, and in rats after a transient cerebral oligemia where very small doses of iron are sufficient to cause apomorphine induced rotations. It is possible, that consequences of mild to moderate transient or permanent local oxygen deficits together with iron might give rise to progressive damage of cerebral tissue in vulnerable structures, effects which are comparable to the neurotoxicity of 6-OHDA.

Neurotoxic effects can also be independently induced by excitotoxic mechanisms. It is known that the mammalian neostriatum is one of the most vulnerable brain structures to cerebral ischemia (Pulsinelli et al., 1982; Ginsberg et al., 1985). Massive release of excitatory amino acids has been suggested to play a major role in the mediation of ischemic cell damage (Rothman and Olney, 1986). During ischemia oxygen free radicals can induce lipid peroxidation which is suggested to be reponsible for injury of the central nervous system (Braughler and Hall, 1989; Hall and Braughler, 1989; Siesjö, 1989). It is also known that a massive increase of extracellular DA concentrations could be observed in rat striatal dialysates subsequently to four vessel occlusion (Globus et al., 1988; Phebus and Clemens, 1989; Obrenovich et al., 1990; Damsma et al., 1990; Akiyama et al., 1991). The four vessel occlusion procedure has been shown to cause damage to striatal neurons (Pulsinelli et al., 1982; Persson et al., 1989).

DA involvement in ischemia-induced neuronal damage has been shown indirectly by the fact that depletion of striatal dopamine protects striatal neurons from ischemia-induced injury (Weinberger et al., 1985; Clemens and Phebus, 1988). Free oxygen radical production during the oxidative deamination of dopamine by monoamine oxydase and auto-oxidation in the basal ganglia is therefore probably involved. Normally, hydrogen peroxide is metabolized mainly by peroxidase. In the presence of transitional metal ions, however, it can be converted to free hydroxyl (OH) radicals (Halliwell and Gutteridge, 1984, 1985). It may be that the increased dopamine release during ischemia in the presence of iron ions is responsible for damaging dopaminergic terminals due to lipid peroxidation.

This is supported by the experiments where transient cerebral oligemia (BCCA procedure) leads to a higher sensitivity to intrastriatally injected iron salts. Also the increased dopamine release following oligemia is comparable to that observed during forebrain ischemia. The release of dopamine is under the control of glutamate (Läer et al., 1992, 1993; Sontag et al., 1992). This oligemic event leads to lipid peroxidation in the frontal cortex and striatum (Melzacka et al., 1992; Sontag et al., 1992).

These data suggest that a minuscule amount of FeCl₃ injected into the ventrolateral striatum is causing the decomposition of preformed lipid peroxides in cerebral structures of old rats which suffer from the consequences of transiently or permanently diminished cerebral partial oxygen pressure after the BCCA treatment. The added metal ions are doing no more than "stimulating" further lipid peroxidation (see Halliwell and Gutteridge, 1989).

How can the behaviour abnormalities following an insult to the ventrolateral striatum be explained? The striatum being a part of the basal ganglia collects signals from a large part of the neocortex, redistributes these cortical inputs both with respect to one another and with respect to inputs from the limbic system, and then focuses the outputs of these redistributed, integrated signals to particular regions of the frontal lobes and brain stem involved in aspects of motor planning and motor memory (Graybiel, 1990). The limbic system partly splits its projections to the striatum into a hippocampus-centred system that focusses on the matrix and an amygdala-centred system that favours striosomes (Ragsdale and Graybiel, 1990); this latter pathway seems to be destroyed in the present experiments. Lesions of the ventrolateral striatum of the rat cause cell loss in the ipsilateral amygdala or piriform cortex (Dunnett and Iversen, 1982).

It may be postulated, therefore, that the cognitive deficiencies of rats lesioned with 6-OHDA and additionally injected with FeCl₃ are due to the interruption of limbic information necessary for the dynamic modulation of behaviour based on sensorimotor memory related and conditional cues derived from the neocortex and limbic system (see also Graybiel, 1990) especially by lesioned afferents of the amygdala and piriform cortex. These pathways are more affected than afferents from CA1 and CA2/CA3 border of the ipsilateral hippocampus (Dunnett and Iversen, 1982).

The Whishaw-test in the water maze seems to test the intactness of the incoming system. The test forces the animals to collect distal cues outside the maze transmitted via an intact afferent system. On the other hand animals are forced not only to collect information but also to switch from one strategy to an other. It was suggested by Cools (Cools, 1981; Cools et al., 1990) that enhanced DAergic activity in the dorsal striatum allows the animals to switch behaviour arbitrarily in contrast to the ventral striatum which allows the animals to switch behaviour with the help of cues. The water maze forces the animals to use cues external to the water tank. It seems therefore that the lesioned ventrolateral striatum prevents limbic informations necessary to be integrated in the striatal function helpful for processing motor function goal directed due to navigation performance.

It is suggested that the lesioning procedure described in the present experiments disrupts information of the ventrolateral striatum necessary for the complete function of the basal ganglia not only for spatial learning and performance and cognitive capacities but also for exploration and spontaneous behaviour.

Acknowledgements

The skilful technical assistance of M. Schindler and H. Ropte is greatly acknowledged. We thank Dr. N.N. Osborne for helpful comments on the manuscript. W. Kolasiewicz and T. Jaros were supported by Bundesministerium für Forschung und Technologie, Förderkennzeichen: 01 KL 9101/0.

References

Abe K, Yuki S, Kogure K (1988) Strong attenuation of ischemic and post ischemic brain edema in rats by a novel free radical scavenger. Stroke 19: 480–485

Akiyama Y, Akihiro I, Koshimura K, Ohue T, Yamagata S, Miwa S, Kikuchi H (1991) Effect of transient forebrain ischemia and reperfusion on function of dopaminergic neurons and dopamine reuptake in vivo rat striatum. Brain Res 561: 120–127

Asano T, Johshita H, Koide T, Takakura K (1984) Amelioration of ischaemic cerebral oedema by a free radical scavenger, AVS: 1.2-bis (nicotinamido)-propane. An experimental study using a regional ischemia model in cats. Neurol Res 6: 163–168

Ben-Schachar D, Youdim MBH (1991) Intranigral iron injection induces behavioural and biochemical "Parkinsonism" in rats. J Neurochem 57: 2133–2135

Betz AL (1990) Effect of the free radical scavenger dimethylthiourea in experimental cerebral ischemia. In: Krieglstein J, Oberpichler H (eds) Pharmacology of cerebral ischemia 1990. Wissenschaftliche Verlagsgesellschaft, Stuttgart, pp 335–342

Block F, Sieklucka M, Schmidt-Kastner R, Heim C, Sontag KH (1993) Metabolic changes during and after transient clamping of carotid arteries in normotensive rats. Brain Res Bull 31: 91–96

Boisvert DPJ, Schreiber C (1992) Interrelationship of excitotoxic and free radical mechanisms. In: Krieglstein J, Oberpichler-Schwenk H (eds) Pharmacology of cerebral ischemia 1992. Wissenschaftliche Verlagsgesellschaft, Stuttgart, pp 311–320

Braughler JM, Hall ED (1989) Central nervous system trauma and stroke. I. Biochemical considerations for oxygen radical formation and lipid peroxidation. Free Rad Biol Med 6: 289–301

Bullock R, Graham DI, Chen MH, Lowe D, McCulloch J (1990) Focal cerebral ischemia in the cat: pretreatment with a competitive NMDA receptor antagonist, D-CPPene. J Cereb Blood Flow Metab 10: 668–674

Cadet JL, Ratz M, Jackson-Lewis V, Fahn S (1989) Vitamine E attenuates the toxic effects of intrastriatal injection of 6-hydroxydopamine (6-OHDA) in rats: behavioural and biochemical evidence. Brain Res 476: 10–15

Clemens JA, Phebus LA (1988) Dopamine depletion protects striatal neurons from ischemia-induced cell death. Life Sci 42: 707–713

Cohen G (1984) Oxyradical toxicity in catecholamine neurons. Neurotoxicology 5: 82–88

Cohen G, Heikkila RE (1974) The generation of hydrogen peroxide, superoxide radical, and hydroradical by 6-hydroxydopamine, dialuric acid, and related cytotoxic agents. J Biol Chem 249: 2447–2452

Cohen G, Heikkila RE, Allis B, Cabbat F, Demblec D, McNamee D, Mytilineau C, Winston B (1976) Destruction of sympathetic nerve terminals by 6-hydroxydopamine: protection by 1-phenyl-3-(2-thiaxolyl)-2-thiourea, diethyldithiocarbamate, methimazole, cysteamine, ethanol, and n-butanol. J Pharmacol Exp Ther 199: 336–352

Cools AR (1981) Aspects and prospects of the concept of neurochemical and cerebral organisation of aggression: introduction of new research strategies in "brain and behaviour" studies. In: Brain PF, Bendton D (eds) Biology of aggression. Fythoff and Noordhoff

Cools AR, Brachten R, Heeren D, Willeman A, Ellenbroek BJ (1990) Search after neurobiological profile of individual-specific features of Wistar rats. Brain Res Bull 24: 49–69

Damsma G, Boisvert DP, Mudrick LA, Wenkstern D, Fibiger HC (1990) Effects of transient forebrain ischemia and pargyline on extra-cellular concentrations of dopamine, serotonine, and their metabolites in the rat striatum as determined by in vivo microdialysis. J Neurochem 54: 801–808

Diemer NH, Balchen T, Hu P, Frank L, Bruhn T, Berg M, Christensen T, Jorgensen MB, Johansen FF (1992) The effect of AMPA anagonists on the regional neuron loss after cerebral ischemia in the rat. In: Krieglstein J, Oberpichler-Schwenk H (eds) Pharmacology of cerebral ischemia. Wissenschaftliche Verlagsgesellschaft, Stuttgart, pp 121–127

Dunnet SB, Iversen SD (1982) Spontaneous and drug-induced rotation following localized 6-hydroxydopamine and kainic acid-induced lesions of the neostriatum. Neuropharmacology 22: 899–908

Flamm ES, Demopoulos HB, Seligman ML, Poser R, Pietronigro DD, Ransohoff J (1978) Free radicals in cerebral ischemia. Stroke 9: 445–447

Ginsberg MD, Graham DI, Busto P (1985) Regional glucose utilization and blood flow following grated forebrain ischemia in the rat: correlation with neuropathology. Ann Neurol 18: 470–481

Ginsberg MD, Takagi K, Globus MY-T (1992) Release of neurotransmitters in the cerebral ischemia: relevance to neuronal injury. In: Krieglstein J, Oberpichler-Schwenk H (eds) Pharmacology of cerebral ischemia. Wissenschaftliche Verlagsgesellschaft, Stuttgart, pp 177–189

Globus MYT, Busto R, Dietrich WD, Martinez E, Valdez I, Ginsberg MD (1988) Effect of ischemia on the in vitro release of striatal dopamine, glutamate, and gamma-aminobutyric acid studied by intracerebral microdialysis. J Neurochem 51: 1455–1464

Graybiel A (1990) Neurotransmitters and neuromodulators in the basal ganglia. TINS 13: 244–254

Haba K, Ogawa N, Mizukawa K, Mori A (1991) Time course of changes in lipid peroxidation, pre- and postsynaptic cholinergic indices, NMDA receptor binding and neuronal death in the gerbil hippocampus following transient ischemia. Brain Res 540: 116–120

Hall ED (1990) Lazaroids. Efficacy and anti-oxidant mechanism in experimental cerebral ischemia. In: Krieglstein J, Oberpichler H (eds) Pharmacology of cerebral ischemia 1990. Wissenschaftliche Verlagsgesellschaft, Stuttgart, pp 343–350

Hall ED, Yonkers PA (1988) Attenuation of postischemic cerebral hypoperfusion by the 21-aminosteroid U74006F. Stroke 19: 340–344

Hall ED, Braughler JM (1989) Central nervous system trauma and stroke. II. Physiological and pharmacological evidence for the involvement of oxigen free radicals and lipid peroxidation. Free Rad Biol Med 6: 303–313

Halliwell B (1978) Biochemical mechanisms accounting for the toxic action of oxygen in living organisms: the key role of superoxid dismutase. Cell Biol Int Rep 2: 113–128

Halliwell B, Gutteridge JMC (1984) Lipid peroxidation, oxygen radicals, cell damage and antioxidant therapy. Lancet i: 1396

Halliwell B, Gutteridge JMC (1985) Oxygen radicals and the nervous system. Trends Neurosci 8: 22–26

Halliwell B, Gutteridge JMC (1989) Free radicals in biology and medicine, 2nd edn. Oxford University Press, pp 86–276

Ito T, Kawakami M, Yamauchi Y (1986) Effect of allopurinol on ischemia and reperfusion-induced cerebral injury in spontaneously hypertensive rats. Stroke 17: 1284–1287

Jonsson G (1980) Chemical neurotoxins as denervation tools in neurobiology. Ann Rev Neurosci 3: 169–187

Kirsch JR, Phelan AM, Lange DG, Traystman RJ (1987) Reperfusion induced free radical formation following global ischemia. Ped Res 21: 202A

König J, Klippel R (1963) The rat brain. A stereotaxic atlas of the forebrain and lower parts of the brain stem. Williams and Wilkins, Baltimore

Kogure K, Arai H, Abe K, Nakano M (1985) Free radical damage of the brain following ischemia. Prog Brain Res 63: 237–259

Kurumaji A, Nehls DG, Park CK, McCulloch J (1989) Effects of NMDA antagonists, MK-801 and CPP, upon local cerebral glucose use. Brain Res 496: 268–284

Läer S, Hüther G, Block F, Sontag KH (1992) Acute reduction of cerebral blood flow by clamping of both carotid arteries increases dopamine release. 4th International Symposium on Pharmacology of Cerebral Ischemia, Marburg, p 67 (Abstract)

Läer S, Block F, Hüther G, Heim C, Sontag KH (1993) Effect of transient reduction of cerebral blood flow in normotensive rats on striatal dopamine release. J Neural Transm [Gen Sect] 92: 203–211

Lekieffre D, Ghribi O, Callebert J, Allix M, Plotkine M, Boulu RG (1992) Effect of kynurenic acid on ischemia-induced glutamate accumulation in the hippocampus during four-vessel occlusion. In: Krieglstein J, Oberpichler-Schwenk H (eds) Pharmacology of cerebral ischemia 1992. Wissenschaftliche Verlagsgesellschaft, Stuttgart, pp 105–112

Lewin G (1985) Clinical trial for Parkinson's disease. Science 230: 527–528

Liu TH, Beckman JS, Freeman BA, Hogan EL, Hsu CY (1989) Polyethylene glycol-conjugated superoxide dismutase and catalase reduce ischemic brain injury. Am J Physiol 256: H589–H593

Lippert K, Welsch M, Krieglstein J (1992) The neuroprotective effect of combined treatment with dizocilpine and NBQX in vitro and in vivo. In: Krieglstein J, Oberpichler-Schwenk H (eds) Pharmacology of cerebral ischemia 1992. Wissenschaftliche Verlagsgesellschaft, Stuttgart, pp 147–153

Martz D, Rayos G, Schielke GP, Betz AL (1989) Allopurinol and dimethylthiourea reduce brain infarction following middle cerebral artery occlusion in the rat. Stroke 20: 488–494

Meldrum BS (1990) Protection against ischemic neuronal damage by drugs acting on excitatory neurotransmission. Cerebrovasc Brain Metabol Rev 2: 27–57

Meldrum BS, Moncada C, Lekieffre D, Arvin B, Smith S (1992) Strategies for cerebraprotection: post-synaptic glutamate antagonism versus inhibition of ischemia-induced glutamate release. In: Krieglstein J, Oberpichler-Schwenk H (eds) Pharmacology of cerebral ischemia 1992. Wissenschaftliche Verlagsgesellschaft, Stuttgart, pp 115–119

Melzacka M, Weiner N, Heim C, Sontag RH, Wesemann W (1992) Effect of transient reduction of cerebral blood flow on membrane anisotropy and lipid peroxidation in different brain areas. 4th International Symposium on Pharmacology of Cerebral Ischemia, Marburg, p 31 (Abstract)

Moncada C, Arvin B, Lekieffre D, Chapman A, Meldrum BS (1992) The non-NMDA anagonist GYKI 52466 inhibits glutamate release induced by transient severe forebrain ischemia in rat striatum. In: Krieglstein J, Oberpichler-Schwenk H (eds) Pharmacology of cerebral ischemia 1992. Wissenschaftliche Verlagsgesellschaft, Stuttgart, pp 155–159

Obrenovitch TP, Sarna GS, Matsunoto T, Symon L (1990) Extracellular striatal dopamine and its metabolites during transient cerebral ischemia. J Neurochem 54: 1526–1532

Oh SM, Betz AL (1992) Interaction between radicals and excitatory amino acids in the formation of ischemic brain edema in rats. Stroke 22: 915–921

Paxinos G, Watson C (1982) The rat brain in stereotaxic coordinates. Academic Press, New York

Persson L, Bolander H, Hillered L, Hardemark HG, Olsson Y, Ponten U (1989) Neurologic and neuropathologic outcome after middle cerebral artery occlusion in rats. Stroke 20: 641–645

Phebus LA, Clemens JA (1989) Effects of transient global, cerebral ischemia on striatal extracellular dopamine, serotonine and their metabolites. Life Sci 19: 1335–1342

Pulsinelli WA, Brierley JB, Plum F (1982) Temporal profile of neuronal damage in a model of forebrain ischemia. Ann Neurol 11: 491–498

Ragsdale CW Jr, Graybiel AM (1990) A simple ordering of neo-cortical areas established by the compartmental organization of their striatal projections. Proc Natl Acad Sci USA 87: 6196–6199

Rothman SM, Olney JW (1986) Glutamate and the pathophysiology of hypoxic ischemic brain damage. Ann Neurol 19: 105–111

Sengstock GJ, Olanow CW, Dunn AJ, Arendash GW (1992) Iron induces degeneration of nigrostriatal neurons. Brain Res Bull 28: 645–649

Sheardown MJ, Hansen AJ, Eskesen K, Suzdak P, Diemer NH, Honore T (1990) Blockade of AMPA receptors in the CA1 region of the hippocampus prevents ischemia induced cell death. In: Krieglstein J, Oberpichler H (eds) Pharmacology of cerebral ischemia 1990. Wissenschaftliche Verlagsgesellschaft, Stuttgart, pp 245–253

Sheardown MJ, Nielsen EO, Hansen AJ, Jacobsen P, Honore T (1990) 2,3-Dihydroxy-6-nitro-7-sulfamyl-benzo(F)quinoxaline: a neuroprotectant for cerebral ischemia. Science 247: 571–574

Siesjö BS (1981) Cell damage in the brain: a speculative synthesis. J Cereb Blood Flow Metab 1: 155–158

Siesjö BK (1989) Free radicals and brain damage. Cerebrovasc Brain Metab Rev 1: 165–211

Siesjö BK, Katsura K, Pahlmark K, Smith ML (1992) The multiple causes of ischemic brain damage: a speculative synthesis. In: Krieglstein J, Oberpichler-Schwenk H (eds) Pharmacology of cerebral ischemia. Wissenschaftliche Verlagsgesellschaft, Stuttgart, pp 511–525

Sontag KH, Heim C, Block F, Sieklucka M, Schmidt-Kastner R, Melzacka M, Osborne N, Läer S, Hüther G, Kunkel M, Ulrich F, Bortolotto Z, Weiner N,

Wesemann W (1992) Cerebral oligemic hypoxia and forebrain ischemia – common and different long-lasting consequences. In: Krieglstein J, Oberpichler-Schwenk H (eds) Pharmacology of cerebral ischemia 1992. Wissenschaftliche Verlagsgesellschaft, Stuttgart, pp 471–479

Ungerstedt U (1971) Postsynaptic supersensitivity after 6-hydroxydopamine-induced degeneration of nigrostriatal dopamine system. Acta Physiol Scand [Suppl] 367: 69–93

Wahl F, Allix M, Plotkine M, Boulu RG (1992) Riluzole reduces infarct size induced by focal cerebral ischemia. In: Krieglstein J, Oberpichler-Schwenk H (eds) Pharmacology of cerebral ischemia 1992. Wissenschaftliche Verlagsgesellschaft, Stuttgart, pp 167–174

Watson BD, Ginsberg MD (1989) Ischemic injury in brain: role of oxygen radical-mediated processes. In: Barkai A, Baran NG (eds) Arachidonic acid in the nervous system. Physiological and pathological significance. Ann NY Acad Sci 559: 269–281

Weinberger J, Nieves-Rosa J, Cohen G (1985) Nerve terminal damage in cerebral ischemia: protective effect of alpha-methyl-para-tyrosine. Stroke 16: 864–870

Whishaw IQ (1987) Hippocampal granule cell and CA 3–4 lesions impair formation of a place learning set in the rat and induce reflex epilepsy. Behav Brain Res 24: 59–72

Whishaw IQ, Dunnett SB (1985) Dopamine depletion, stimulation or blockade in the rat disrupts spatial navigation and locomotion dependent upon beacon or distal cues. Behav Brain Res 18: 11–29

Yamamoto M, Shima T, Uozumi T, Sogabe T, Yamada K, Kawasaki T (1983) A possible role of lipid peroxidation in cellular damages caused by cerebral ischemia and the protective effect of tocopherol administration. Stroke 14: 977–982

Yoshida S, Abe K, Busto R, Watson BD, Kogure K, Ginsberg MD (1982) Influence of transient ischemia on lipid-soluble antioxidants, free fatty acids and energy metabolites in rat brain. Brain Res 245: 307–316

Young W, Wojak JC, DeCresito V (1988) 21-Aminosteroid reduces ion shifts and edema in the rat middle cerebral artery occlusion model of regional ischemia. Stroke 19: 1013–1019

Correspondence: Dr. C. Heim, Department of Psychiatry, University of Göttingen, von-Siebold-Strasse 5, D-37075 Göttingen, Federal Republic of Germany.

Cytokine induced synthesis of nitric oxide from L-arginine: a cytotoxic mechanism that targets intracellular iron

J. B. Hibbs, Jr.

VA Medical Center and Division of Infectious Diseases, School of Medicine, The University of Utah, Salt Lake City, Utah, U.S.A.

Summary

A cytokine inducible high output nitric oxide synthase was recently identified. It is induced by cytokines in macrophages as well as in nonmacrophage cell types. It is a product of the cell-mediated immune response and probably has multiple functional roles. Current experimental results suggest that cytokine induced synthesis of nitric oxide from L-arginine has a role in the defense of the intracellular environment against intracellular microbes. Nitrosylation of intracellular iron, particularly nonheme iron, appears to be a major biochemical fate of nitric oxide synthesized by the cytokine inducible nitric oxide synthase. This results in inhibition of redox enzymes that have nonheme iron essential for catalytic activity. Because nitric oxide is paramagnetic (has an unpaired electron) it can readily react with dioxygen and the superoxide anion to yield toxic products. As a result, the balance between beneficial and deleterious redox reactions involving nitric oxide must be tightly regulated by currently unidentified mechanisms.

Introduction

It has been recognized for many years that cooperation of macrophages and T-lymphocytes are necessary for the development of cell-mediated immune (CMI) responses (Hibbs et al., 1990). It also was recognized that the CMI response is important in host defense against intracellular microorganisms (Mackaness, 1971; Hibbs et al., 1980). What was not known, however, was the identity of biochemical defenses, induced by the CMI response, that attacked microorganisms that had invaded the intracellular environment. Earlier experiments had demonstrated that

cytokines generated during CMI responses dramatically modified macrophage function (Mackaness, 1971; Hibbs et al., 1980, 1990). Macrophages removed from active CMI reactions in vivo or macrophages
treated with cytokine containing supernatants plus lipopolysaccharide
(LPS) or interferon-gamma (IFN-γ) plus LPS in vitro were cytotoxic.
Immunologically activated macrophages interestingly acquired the ability to express nonspecific cytotoxicity not only for intracellular pathogens,
but also for neoplastic cells (Hibbs et al., 1971, 1972, 1977, 1980, 1990;
Weinberg et al., 1978). We now know that cytokine induced synthesis of

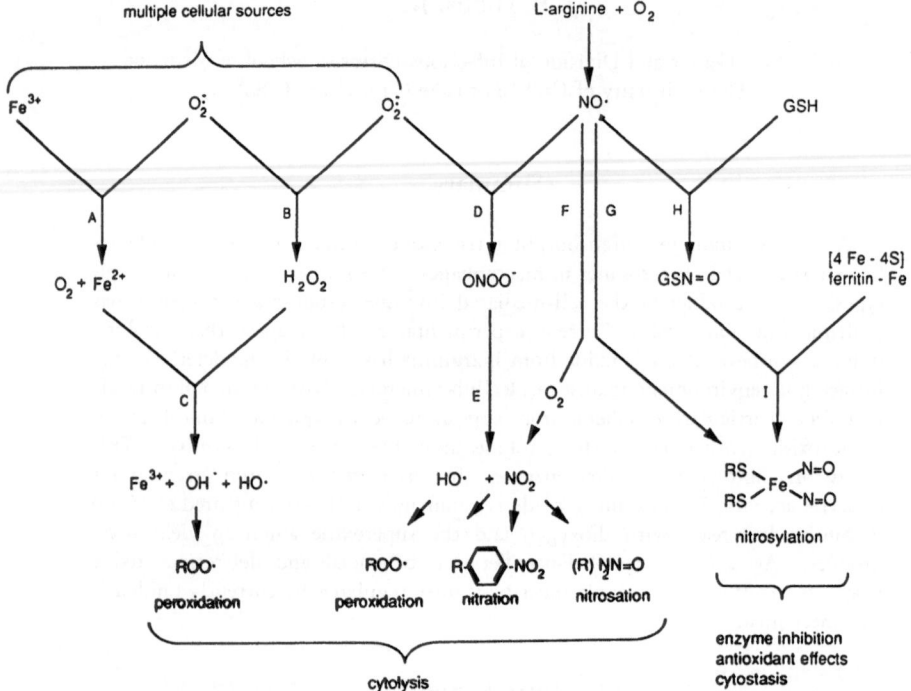

Fig. 1. Schematic representation of pathways synthesizing low-molecular-weight reactive
molecules active in host defence. We have arbitrarily termed addition of the nitroso group
to iron and thiols as nitrosyiation and to other atoms as nitrosation. Additional products of
pathway F include N_2O_3, N_2O_4, NO_2, and NO_3 (Hibbs et al., 1990; Leaf et al., 1990). GSH
may not react directly with NO flowing through pathway H, but with NO metabolites such
as N_2O_3 and N_2O_4 (unpaired electrons are symbolized by ˙). Abbreviations used include:
nitrogen dioxide (NO_2˙), reduced glutathione (GSH), S-nitrosoglutathione (GSN = O), and
paramagnetic mononuclear iron sulfur dinitrosyl complexes $[(RS)_2Fe(NO˙)_2]^-$. Heme nitrosyl
complexes are also formed by pathways G, H, and I (Bastian et al., 1992; Hibbs et al.,
1992a), but are not shown. Abbreviations used include: nitrite (NO_2^-) and nitrate (NO_3^-).
[From Hibbs (1992) with permission of the publisher]

nitric oxide (NO˙) from L-arginine is an important effector molecule in mediating host resistance to intracellular pathogens and that NO˙ has a role (but less well defined) in host resistance to neoplastic cells (Hibbs et al., 1990; Nathan and Hibbs, 1991).

Because NO˙ is paramagnetic and redox active, it reacts readily with cellular iron and with O_2 in its several redox forms (see Fig. 1) (Beckman et al., 1990; Hibbs, 1992; Stamler et al., 1992). In our view, the iron and thiol nitrosylation pathways have biochemical characteristics needed for a viable defense of the intracellular environment against microbes. Iron and thiol nitrosylation reactions could control intracellular proliferation of intracellular pathogens while causing minimal collateral damage to host cell molecules and organelles (Hibbs, 1992). However, it is obvious that redox reactions of NO˙ with O_2 and the superoxide anion (O_2^-) can readily occur. The reaction of NO˙ with O_2 and O_2^- can produce very toxic products capable of causing indiscriminate damage in the tissues (Beckman et al., 1990; Hibbs, 1992; Wink et al., 1992). Therefore, once the cytokine induced high output NO synthase (iNOS) is activated, NO˙ must flow into the iron nitrosylation pathway if indiscriminate collateral damage to host tissue is to be minimized (Hibbs, 1992). Our experimental results, as well as those of other laboratories, suggest that iron nitrosylation is an important fate of cytokine induced NO˙ in many cell types (Lancaster and Hibbs, 1990; Pellat et al., 1990; Drapier et al., 1991; Bastian et al., 1992; Hibbs et al., 1992; Lancaster et al., 1992). However, redox reactions of NO with O_2 and O_2^- must occur in many pathophysiological situations (see Fig. 1). In this case, the potential for severe cytotoxic damage to autologous tissue is high because of the potent peroxidizing, nitrating, and nitrosating species such as hydroxyl radical (HO˙) (Halliwell and Gutteridge, 1990), nitrogen dioxide (NO_2) (Prütz et al., 1985), nitrogen trioxide (N_2O_3) (Leaf et al., 1990), as well as other highly reactive oxynitrogen intermediates that can be formed (Wink et al., 1992)

The following is a brief review of our laboratory's contribution to the discovery of the synthesis of NO˙ from L-arginine. The results demonstrate the importance of iron nitrosylation reactions in the function of NO˙ as an effector molecule of CMI responses.

It was known that the cytokine inducible activated macrophages cytotoxic effector mechanism targeted enzymes with catalytically essential nonheme iron before the biological synthesis of NO˙ from L-arginine was discovered

Target cells co-cultivated with cytotoxic activated macrophages develop inhibition of mitochondrial respiration but continue to synthesize

ATP via the glycolytic pathway (Granger et al., 1980). This effect is due
to inhibition of the proximal two oxidoreductases of the mitochondrial
electron transport system NADH: ubiquinone oxidoreductase (Complex
I) and succinate: ubiquinone oxidoreductase (Complex II) (Granger and
Lehninger, 1982; Hibbs et al., 1984; Drapier and Hibbs, 1986, 1988)
(see Fig. 2 for a schematic representation of the citric acid cycle and the
electron transport chain). Other enzymes of the mitochondrial electron
transport system of tumor target cells are not affected. Glycolysis, a
metabolic pathway which does not have catalytically active iron, remains
functional in target cells of activated macrophages. Complex I and
Complex II are iron-sulfur [4Fe-4S] flavoprotein dehydrogenases that
contain all the [4Fe-4S] complexes of the mitochondrial electron transport chain (Complex III has a [2Fe-2S] cluster). The activated macrophage effector cells develop the same pattern of metabolic inhibition as
their tumor target cells (Drapier and Hibbs, 1988).

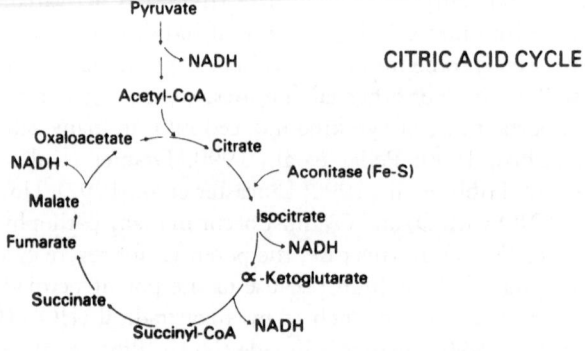

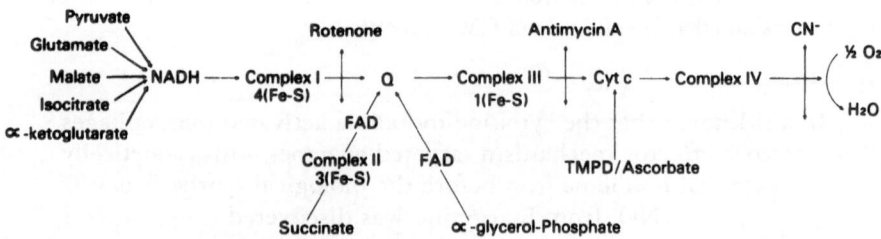

Fig. 2. Schematic representation of the citric acid cycle and the electron transport chain
[From Drapier and Hibbs, 1986, with permission of the publisher]

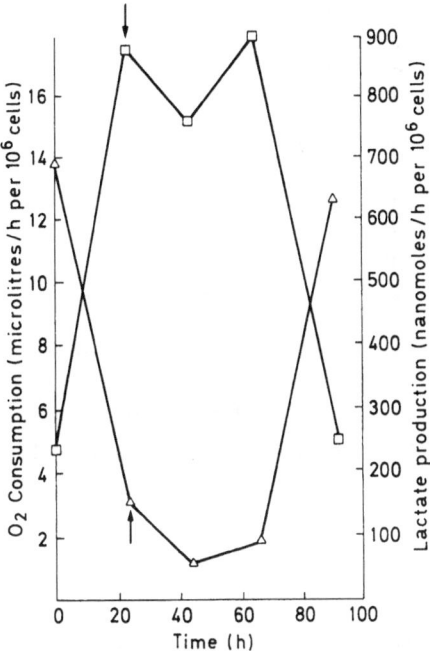

Fig. 3. Respiration and glycolysis of L1210 cells co-cultivated with CAM. At the times shown L1210 cells were removed from culture and the rates of dinitrophenol-uncoupled respiration (△) and aerobic glycolysis (□) were measured. The arrows show the time at which L1210 cells were separated from CAM and reincubated in fresh culture medium [From Hibbs and Granger (1982) with permission of the publisher]

Figure 3 shows the correlation between mitochondrial respiration and glycolysis following a 21-hour period of co-cultivation of L1210 cells with cytotoxic activated macrophages (Hibbs and Granger, 1982). After removal from the monolayer of cytotoxic activated macrophages (the arrows show when the L1210 cells were separated from the macrophages), the cytostatic L1210 cells were cultured in the absence of macrophages in fresh medium with daily replenishment. At the intervals shown, aliquots of the population of L1210 cells were removed from culture, washed by centrifugation, and respiration and aerobic glycolysis were measured. During the initial 21 hour co-culture period, respiration fell to a low level, and during the same time, the rate of glycolysis rose by more than three fold. During subsequent culture in the absence of macrophages, while the L1210 cell remained non-dividing (hours 21–68), respiration was inhibited and glycolysis was elevated. However, once cell division resumed at about 90 hours, respiration had risen to the control rate and aerobic glycolysis had fallen to the control level. The recovery of L1210

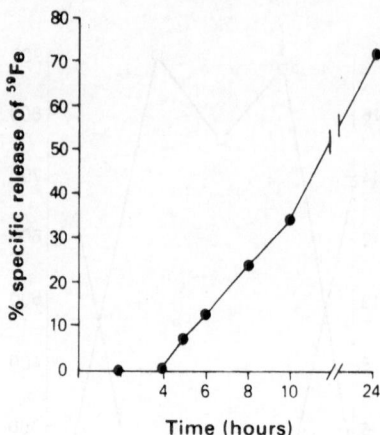

Fig. 4. Kinetics of ^{59}Fe specific release from viable murine L1210 leukemia cells co-cultivated with CAM [Adapted from Hibbs et al. (1984) with permission of the publisher]

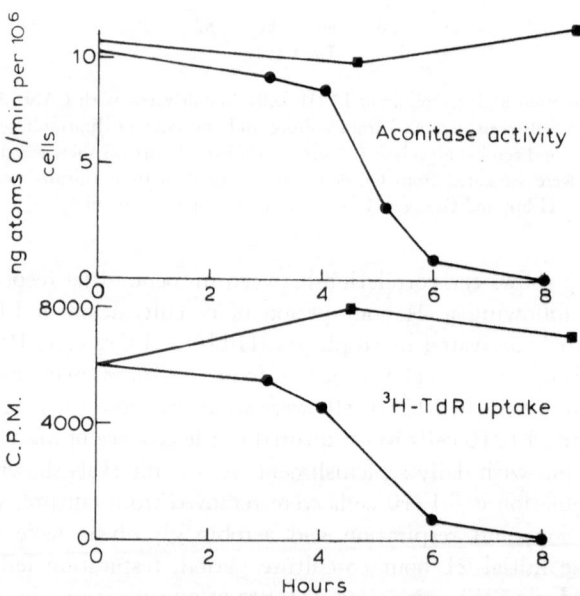

Fig. 5. The kinetics of citrate-dependent respiration (aconitase activity) and (^{3}H) thymidine incorporation into DNA in L10 cells cultured alone (■) or co-cultivated with cytotoxic activated macrophages (●). Measurements of aconitase activity and DNA synthesis were as described [From Drapier and Hibbs (1986) with permission of the publisher]

cell division was associated with a return of L1210 cell energy metabolism to the normal pattern. The L1210 cells remained viable throughout the duration of the experiment. Resumption of mitochondrial respiration (and DNA replication) was due to recovery of the entire population of injured L1210 cells and not outgrowth of a small number of unaffected cells as determined by cloning injured L1210 cells in soft agar (Hibbs and Granger, 1982).

Figure 4 demonstrates that cytotoxic activated macrophages cause a major perturbation of iron homeostasis in tumor target cells. Specific release of iron label from target cells began 4–6 hour after initiating cocultivation (Hibbs et al., 1984). This was the time point that inhibition of DNA synthesis and aconitase activity was first detected (Drapier and Hibbs, 1986) (see Fig. 5).

Cytotoxic activated macrophages induce inhibition of DNA replication as determined by (^{3}H) thymidine uptake by tumor target cells (Granger et al., 1980). Ribonucleotide reductase, the rate limiting enzyme in DNA synthesis, contains nonheme iron essential for catalytic activity. Recent experiments show that inhibition of ribonucleotide reductase is mediated by NO˙ (Lepoivre et al., 1990; Kwon et al., 1991).

Aconitase, a citric acid cycle enzyme with a catalytically active [4Fe-4S] prosthetic group is reversibly inhibited in target cells of cytotoxic activated macrophages and in the macrophages themselves (Drapier and Hibbs, 1986, 1988) (see Fig. 5). A causal relationship between removal of iron from the [4Fe-4S] group of aconitase and inhibition of aconitase activity was established because enzymatic activity was reconstituted by incubation with ferrous ions or ferrous ions plus a reducing agent (Drapier and Hibbs, 1986).

Inhibition of enzymes with nonheme iron essential for catalytic activity is due to cytokine induced synthesis of NO˙ from L-arginine

In 1987, we observed that L-arginine is required for expression of the activated macrophage cytotoxic effector mechanism that causes inhibition of mitochondrial respiration, aconitase activity, and DNA synthesis in tumor target cells (Hibbs et al., 1987a). Cytotoxic activated macrophages synthesize NO˙ from a terminal guanidino nitrogen atom of L-arginine which is converted to L-citrulline without loss of the guanidino carbon atom (Hibbs et al., 1987b, 1988; Iyengar et al., 1987; Marletta et al., 1988) (see Fig. 6). Authentic NO˙ gas caused the same pattern of cytotoxicity in L10 hepatoma cells as is induced by cytotoxic activated macrophages (iron loss as well as inhibition of DNA synthesis, mitochondrial respiration, and aconitase activity) (Hibbs et al., 1988; Stuehr and Nathan, 1989). These results demonstrate that NO˙ is the precursor of the nitrite/nitrate synthesized by cytotoxic activated macrophages and, suggested that, via formation of iron nitrosyt complexes, an effector molecule. Figure 7 shows that the results of an electron paramagnetic resonance spectroscopy study performed in cytotoxic activated macrophage effector cells, which develop the same pattern of metabolic inhibition as their targets (Lancaster and Hibbs, 1990). Examination of activat-

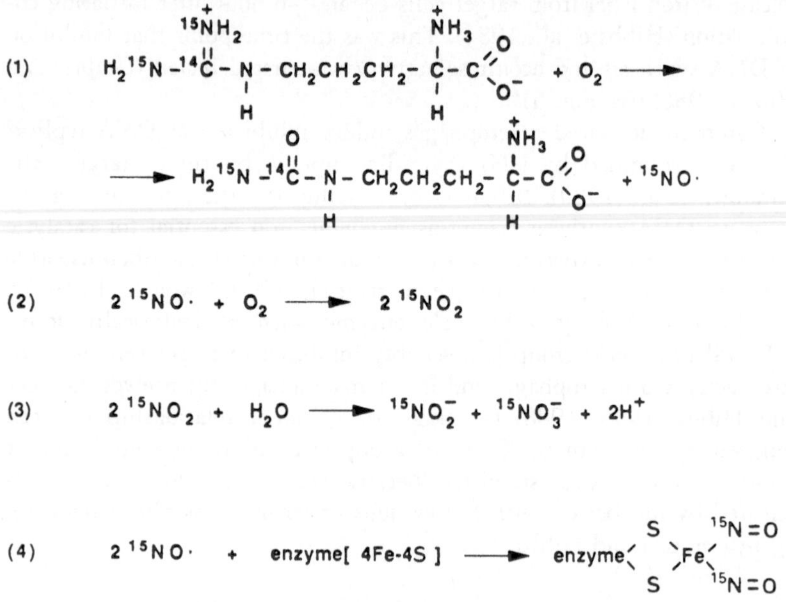

Fig. 6. Precursor and products of the biological synthesis of inorganic nitrogen oxides and L-citrulline from L-arginine. ^{15}N-containing products derived from L-[guanidino-$^{15}N_2$]-arginine were identified by gas chromatography/mass spectrometry (Iyengar et al., 1987; Hibbs et al., 1988; Marletta et al., 1988; Palmer et al., 1988) or electron paramagnetic resonance spectroscopy (Lancaster and Hibbs, 1990) except for nitrogen dioxide (NO_2), which was detected by another method (Hibbs et al., 1988). The direct synthesis of L-citrulline from L-arginine has been identified with several techniques (Hibbs et al., 1987a, 1988; Iyengar et al., 1987; Amber et al., 1988). The experiments utilizing L-[guanidino-^{14}C]-arginine (Amber et al., 1988; Hibbs et al., 1988) are illustrated in the figure. NO˙ formed by reaction (1) can under go oxidative degradation in aqueous solution [reactions (2) and (3) or react with non-heme iron associated with sulfur atoms to form nitrosyl-iron-sulfur complexes (reaction 4)]. Although not shown, certain other forms of intracellular iron also probably complex with NO˙. NO_2^- that contacts oxyhemoglobin is rapidly oxidized to NO_3^- (Kosaka et al., 1979). Therefore, NO˙ synthesized from L-arginine by either the cytokine induced high output pathway or the constitutive low output pathway will be detected in serum or urine as NO_3^-. Abbreviations used include: nitrite (NO_2^-) and nitrate (NO_3^-)
[Modified from Hibbs et al. (1990) with permission of the publisher]

ed macrophages from mice infected with *Mycobacterium bovis* (strain bacillus Calmette-Guérin) that were cultured in medium with LPS and L-arginine showed the presence of an axial signal at g = 2.039, which is similar to previously described iron-nitrosyl complexes formed form the destruction of iron-sulfur centers by authentic NO˙. Inhibition of the L-arginine-dependent pathway by addition of N^{ω}-monomethyl-L-arginine (methyl group on a terminal guanidino nitrogen and which had previous-ly been shown to be a potent inhibitor of the inducible NO synthase) (Hibbs et al., 1987a, b) decreased the production of nitrite, nitrate, and the g = 2.039 signal. Comparison of the hyperfine structure of the signal

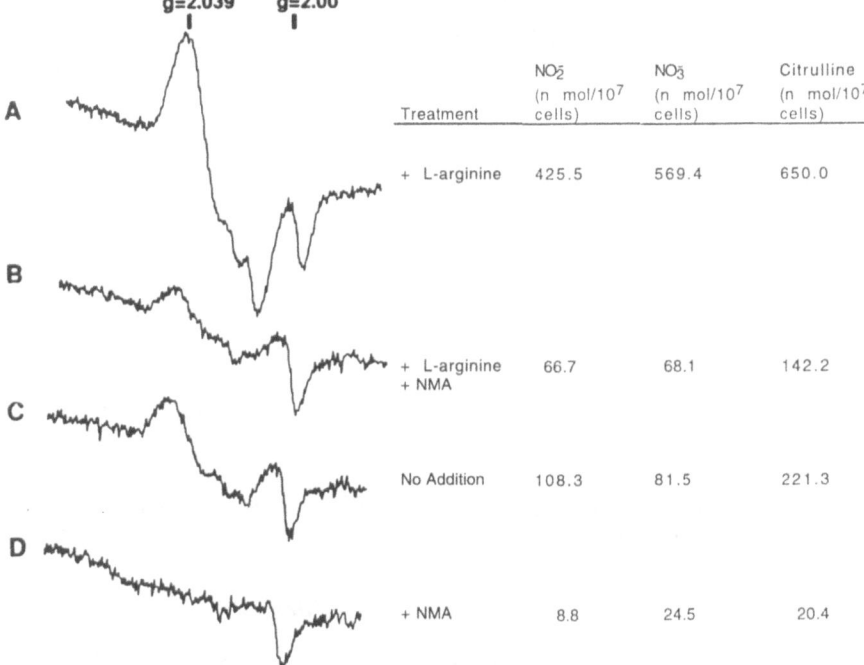

Treatment	NO_2^- (n mol/10^7 cells)	NO_3^- (n mol/10^7 cells)	Citrulline (n mol/10^7 cells)
+ L-arginine	425.5	569.4	650.0
+ L-arginine + NMA	66.7	68.1	142.2
No Addition	108.3	81.5	221.3
+ NMA	8.8	24.5	20.4

Fig. 7. Simultaneous synthesis of NO_2^-, NO_3^-, and citrulline by cytotoxic activated macro-phages (CAM) and development of the g = 2.039 signal characteristic of $[NO\text{-}Fe\text{-}S]^-$ complexes in the cytoplasm of CAM. Peritoneal macrophages activated in vivo by infectious with BCG were cultured for 20.5 h in medium containing LPS 20 ng/ml and prepared without amino acids (basic medium). The CAM were cultured in basic medium without further modification (A) or with the additives indicated in the figure. The NO_2^-, NO_3^-, and L-citrulline synthesized were then measured and the CAM removed and frozen at –80 °C. Whole CAM were examined by EPR spectroscopy with modulation amplitude 10 gauss and relative instrument gain 1.6 × 10^3. Abbreviations used include: nitrite (NO_2^-), nitrate (NO_3^-), and N^{ω}-monomethyl-L-arginine (NMA) [From Lancaster and Hibbs (1990) with permission of the publisher]

from cells treated with L-arginine with terminal guanidino nitrogen atoms of natural abundance N^{14} atoms or labeled with N^{15} atoms showed that the nitrosyl group in this paramagnetic species arises from one of these two atoms (Lancaster and Hibbs, 1990). These results show that loss of iron-containing enzyme function in cytotoxic activated macrophages (Drapier and Hibbs, 1986) is a result of the formation of iron-nitrosyl complexes induced by the synthesis of nitric oxide from the oxidation of a terminal guanidino nitrogen atom of L-arginine.

Induction of iNOS and expression of L-arginine dependent cytotoxicity requires at least two differentiation signals

Macrophages from normal mice are not activated and are not cytotoxic for intracellular pathogens or mammalian cells (see Fig. 8). Acquisition of cytotoxic activity by activated macrophages requires several signals (Hibbs et al., 1977; Weinberg et al., 1978). The first signal produces

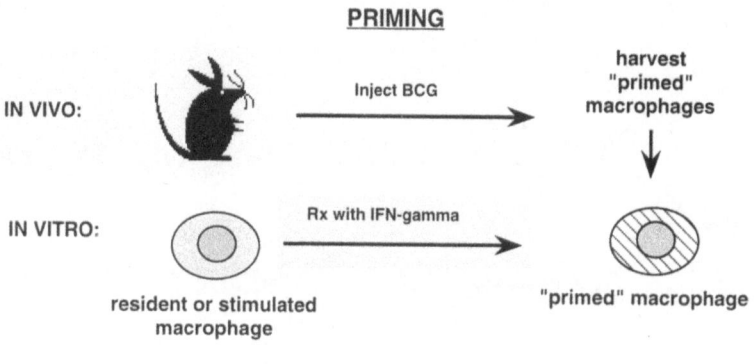

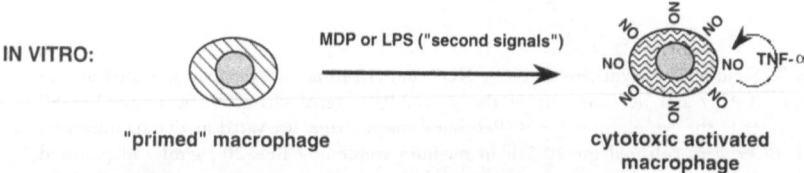

Fig. 8. Induction of the high output NO synthase (iNOS) requires at least two differentiation signals. Resident macrophages and stimulated macrophages (elicited by sterile inflammatory stimulants such as thioglycollate) are not cytotoxic for intracellular pathogens or mammalian cells (Hibbs et al., 1977) when cultured in the absence of differentiation signals (see text for explanation) [From Hibbs et al. (1992b) with permission of the publisher]

primed macrophages or noncytotoxic activated macrophages. Primed macrophages are produced in mice with chronic infection with an intracellular pathogen such as the BCG strain of *Mycobacterium bovis* or by in vitro treatment of peritoneal macrophages form normal mice with interferon-γ (IFN-γ) (Hibbs et al., 1977; Weinberg et al., 1978; Drapier et al., 1988; Ding et al., 1988). However, full induction of the high output L-arginine NO pathway and expression of L-arginine dependent cytotoxicity requires a second step, e.g., exposure to a microbial product such as LPS (Hibbs et al., 1977; Weinberg et al., 1978; Drapier et al., 1988; Ding et al., 1988). Recent studies show that tumor necrosis factor (TNF) is a potent second signal (Drapier et al., 1988; Ding et al., 1988). Microbial products such as muramyl dipeptide (MDP) or LPS act, at least in part by inducing TNF synthesis which is a physiologic cosignal and final inducer of the high output L-arginine: NO pathway in macrophages primed by IFN-γ (Drapier et al., 1988). Therefore, classical cell-mediated immune mechanisms (in the presence or absence of certain microbial components) induce the high output L-arginine: NO pathway in macrophages.

Cytokines induce the high output NO synthase in nonmacrophages cells as well as in macrophages

Cytokines also induce the high output iNOS in nonmacrophage cells (Amber et al., 1988a, b; Hibbs et al., 1990; Amber et al., 1991). We observed that conditioned medium from cultures of cytotoxic activated macrophages induced murine mammary adenocarcinoma EMT-6 cells to synthesize NO˙ (Amber et al., 1988a). We then observed that the same combinations of cytokines and LPS that induced high output NO synthesis by macrophages (see Fig. 8) were equally effective in inducing high output NO˙ synthesis by EMT-6 cells and normal murine fibroblasts (Amber et al., 1988b). If EMT-6 cells or fibroblasts are substituted for resident macrophages depicted in the scheme outlined Fig. 8 the induction of high output NO synthesis occurs in the EMT-6 cells or fibroblasts exactly as illustrated (Amber et al., 1988b). Similar to macrophages, IFN-γ was necessary, but not sufficient, for induction of high output NO˙ synthesis. Likewise, as in macrophages (Drapier et al., 1988; Ding et al., 1988), reagent TNF functioned as a cosignal capable of inducing expressions of high output NO synthesis in IFN-γ primed EMT-6 cells and fibroblasts. However, we noted that interleukin-1 (IL-1), which is not active as a cosignal in the induction of iNOS in macrophages (Drapier et al., 1988; Ding et al., 1988), was a potent cosignal active in iNOS induction in EMT-6 cells and fibroblasts (Amber et al., 1988b). We later isolated and identified IFN-γ and TNF in conditioned medium from cytotoxic activated macrophages (Amber et al., 1991). Treatment of conditioned medium with antibodies to IFN-γ and TNF neutralized its ability to induce high output NO synthesis in EMT-6 cells.

We also observed that when EMT-6 cells are induced to synthesize NO˙ by cytokines that they develop the same pattern of inhibition of enzymes with nonheme iron dependent catalytic activity (Amber et al., 1988a, b) as we observed earlier in activated macrophages and their target cells (Hibbs et al., 1987a, b; Drapier and Hibbs, 1988).

Taken together, these results demonstrated: (1) that conditioned medium generated in vitro from cells normally present in the tissue during CMI reactions contained cytokines known to induce iNOS in macrophages; (2) that these same cytokines readily induce iNOS in adenocarcinoma cells and fibroblasts; and (3) that once iNOS is induced by cytokines in nonmacrophage cells iron nitrosylation reactions occur which causes iron loss and a characteristic pattern of enzymatic inhibition. Subsequent work in other laboratories has shown cytokines can induce iNOS in a number of other nonmacrophage somatic cells. Some examples include hepatocytes (Curran et al., 1989; Chan et al., 1992) and endothelial cells (Kilbourn and Belloni, 1990). It is probable that cytokines can induce iNOS in most nucleated somatic cells.

A major role for iNOS during CMI reactions appears to be defense of the intracellular environment against invasion by faculatative and obligate intracellular pathogens (Hibbs et al., 1990; Nathan and Hibbs, 1991). Cytokine inducible NOS is a cytoplasmic enzyme which has been detected in many somatic cell phenotypes. It synthesizes an effector molecule (NO˙) that causes iron nitrosylation, particularly of nonheme iron. Iron nitrosylation can cause microbial cytostasis and other biochemical effects while producing minimal collateral damage to infected host cells (Hibbs, 1992). Clearly other roles for cytokine induced high output NO˙ synthesis during CMI reactions must exist. One likely effect of cytokine induced NO˙ synthesis would be to match perfusion to the metabolic demands of the CMI reaction. In this case the vascular relaxing properties of NO˙ (Palmer et al., 1988) would function to increase blood flow and hence mobilization of cells and molecules needed for host defense in the local anatomical region in the CMI reaction. Therefore, NO˙ synthesized by iNOS could have a similar function to the much smaller amounts of NO˙ synthesized by the constitutive NOS (cNOS) during normal metabolic activity in many different tissues and organs, e.g. coupling of blood flow to metabolic demand (Gally et al., 1990).

In conclusion, evidence has rapidly accumulated supporting the view that the cytokine inducible NO synthesis is a major molecular defense of the intracellular environment against facultative and obligate intracellular microbes. Cytokine inducible NO˙ has potent cytotoxic effects for the facultative intracellular fungal pathogen *Cryptococcus neoformans* (Granger et al., 1990), intracellular amastigotes of *Leishmania major* (Green et al., 1990; Liew et al., 1990), intracellular trophozoite of *Toxoplasma gondii* (Adams et al., 1990), plasmodia (Nussler et al., 1991), the obligate intracellular bacteria *Chlamydia trachomatis* (Mayer et al., 1993), as well as the facultative intracellular bacteria *Mycobacterium leprae* (Adams et al., 1991), Mycobacterium tuberculosis (Denis, 1977; Chan et al., 1992), *Francisella tularensis* (Fortier et al., 1992), and *Listeria monocytogenes* (Beckerman et al., 1993).

In addition to defending the intracellular environment against intracellular microbes, cytokine induced NO˙ synthesis likely has other important roles during CMI reactions that will be defined by future studies. Unregulated or inappropriate NO˙ synthesis by iNOS or cNOS isoforms could also be the cause of cellular damage in many pathophysiological situations. Inappropriate iron nitrosylation reactions or redox interactions of NO˙ with O_2 or O_2^- can yield highly reactive oxynitrogen species that would be destructive to autologous cells. The production of high output NO˙ synthesis during CMI reactions was obviously selected by evolutionary forces because of beneficial effects that resulted for the host. However, like all redox active low molecular weight molecules, NO˙ is capable of causing toxicity. The balance between benefit and damage may be a delicate one when cytokines induce high output NO˙ synthesis in the tissues.

References

Adams LB, Hibbs JB Jr, Taintor RR, Krahenbuhl JL (1990) Microbiostatic effect of murine macrophages for *Toxoplasma gondii:* role of synthesis of inorganic nitrogen oxides from L-arginine. J Immunol 144: 2725–2729

Adams LA, Franzblau AG, Vavrin Z, Hibbs JB Jr, Krahenbuhl JL (1991) L-arginine-dependent macrophage effector functions inhibit metabolic activity of *Mycobacterium leprae.* J Immunol 147: 1642–1646

Amber IJ, Hibbs JB Jr, Taintor RR, Vavrin Z (1988a) L-arginine dependent effector mechanisms is induced in murine adenocarcinoma cells by culture supernotant from cytotoxic activated macrophages. J Leukoc Biol 43: 187–192

Amber IJ, Hibbs JB Jr, Taintor RR, Vavrin Z (1988b) Cytokines induce an L-arginine-dependent effector system in non-macrophage cells. J Leukoc Biol 44: 58–65

Amber IJ, Hibbs JB Jr., Parker CJ, Johnson BB, Taintor RR, Vavrin Z (1991) Activated macrophage condition medium: identification of the soluble factors inducing cytotoxicity and the L-arginine dependent effector mechanism. J Leukoc Biol 49: 610–620

Bastian NR, Xu S, Shao XL, Shelby J, Hibbs JB Jr (1992) Nitric oxide production in response to allogeneic heart transplant in mice. In: Moncada S, Marletta MA, Hibbs JB Jr, Higgs EA (eds) The biology of nitric oxide, vol 2. Enzymology, biochemistry and immunology. Portland Press, London Chapel Hill, pp 273–276

Beckerman KP, Rogers HW, Corbett JA, Schreiber RD, McDaniel ML, Unanue ER (1993) Release of nitric oxide during the T Cell-independent pathway of macrophage activiation: its role in resistance to Listeria monocytogenes. J Immunol 150: 888–895

Beckman JS, Beckman TW, Chen J, Marshall PA, Freeman BA (1990) Apparent hydroxyl radical production by peroxynitrite: implications for endothelial injury from nitric oxide and superoxide. Proc Natl Acad Sci USA 87: 1620–1624

Chan J, Xing Y, Magliozzo RS, Bloom BR (1992) Killing of virulent Mycobacterium tuberculosis by reactive nitrogen intermediates produced by activated murine macrophages. J Exp Med 175: 1111–1122

Curran RD, Billiar TR, Stuehr DJ, Simmons RL (1989) Hepatocytes produce nitrogen oxides from L-arginine in response to inflammatory products of kupffer cells. J Exp Med 170: 1796–1774

Denis M (1991) Interferon-gamma-treated murine macrophages inhibit growth of tubercle bacilli via the generation of reactive nitrogen intermediates. Cell Immunol 132: 150–157

Ding AH, Nathan CF, Stuehr DJ (1988) Release of reactive nitrogen intermediates and reactive oxygen intermediates from muse peritoneal macrophages. J Immunol 141: 2407–2414

Drapier J-C, Hibbs JB Jr (1986) Murine cytotoxic activated macrophages inhibit aconitase in tumor cells. Inhibition involves the iron-sulfur prosthetic group and is reversible. J Clin Invest 78: 790-797

Drapier J-C, Hibbs JB Jr (1988) Differentiation of murine macrophages to express nonspecific cytotoxicity for tumor cells results in L-arginine-dependent inhibition of mitochondrial iron-sulfur enzymes in the macrophage effect cells. J Immunol 140: 2829–2838

Drapier J-C, Wietzerbin J, Hibbs JB Jr (1988) Interferon-γ and tumor necrosis factor induce the L-arginine-dependent cytotoxic effector mechanism in murine macrophages J Immunol 18: 1587–1592

Drapier J-C, Pellat C, Yann H (1991) Generation of EPR-detectable nitrosyl-iron complexes in tumor target cells cocultured with activated macrophages. J Biol Chem 266: 10162–10167

Fortier AH, Polsinelli T, Green SJ, Nacy CA (1992) Activation of macrophages for destruction of Francisella tularensis: identification of cytokines, effector cells, and effector molecules. Infect Immun 60: 817–825

Gally JA, Montague PR, Reeke GN, Edelman GM (1990) The NO hypothesis: possible effects of a short-lived, rapidly diffusible signal in the development and function of the nervous system. Proc Natl Acad Sci USA 87: 3547–3551

Granger DL, Taintor RR, Cook JL, Hibbs JB Jr (1980) Injury of neoplastic cells by murine macrophages leads to inhibition of mitochondrial respiration. J Clin Invest 65: 357–370

Granger DI, Lehninger Al (1982) Sites of inhibition of mitochondrial electron transport in macrophage-injured neoplastic cells. J Cell Biol 95: 527–535

Granger DL, Hibbs JB Jr, Perfect JR, Durack DT (1990) Metabolic fate of L-arginine in relation to microbiostatic capability of macrophages. J Clin Invest 85: 264–273

Green SJ, Meltzer MS, Hibbs JB Jr, Nacy CA (1990) Activated macrophages destroy intracellular Leishmania major amastigotes by an L-arginine dependent killing mechanism. J Immunol 144: 278–283

Halliwell B, Gutteridge MC (1990) Role of free radicals and catalytic metal ions in human disease: an overview. Methods Enzymol 186: 1–85

Hibbs JB Jr (1992) Overview of cytotoxic mechanisms and defense of the intracellular environment against microbes. In: Moncada S, Marletta MA, Hibbs JB Jr, Higgs EA (eds) The biology of nitric oxide, vol 2. Enzymology, biochemistry and immunology. Portland Press, London Chapel Hill, pp 201–206

Hibbs JB Jr, Granger DL (1982) Activated macrophage-induced cytostasis and inhibition of aerobic energy metabolism in transformed cells: evaluation of lytic and nonlytic target cell responses. In: Mizuno D, Cohn ZA, Takeya K, Ishida N (eds) Self-defence mechanisms. Role of macrophages. University of Tokyo, Tokyo, pp 319–333

Hibbs JB Jr, Lamber LH Jr, Remington RS (1971) Resistance to murine tumors conferred by chronic infection with intracellular protozoa, Toxoplasma gondii and Besnoitia jellisoni. J Infect Dis 124: 587–592

Hibbs JB Jr, Lambert LH Jr, Remington RS (1972) In vitro nonimmunologic destruction of cells with abnormal growth characteristics by adjuvant activated macrophages. Proc Soc Exp Biol Med 139: 1049–1052

Hibbs JB Jr, Taintor RR, Chapman HA Jr, Weinberg JB (1977) Macrophage tumor killing: influence of the local environment. Science 197: 279–282

Hibbs JB Jr, Remington JS, Stewart CC (1980) Modulation of immunity and host resistance by micro-organisms. Pharmacol Ther 8: 37–69

Hibbs JB Jr, Taintor RR, Vavrin Z (1984) Iron depletion: possible cause of tumor cell cytotoxicity induced by activated macrophages. Biochem Biophys Res Commun 123: 716 723

Hibbs JB Jr, Taintor RR, Vavrin Z (1987a) Macrophage cytotoxicity: role for L-arginine deiminase activity and imino nitrogen oxidation to nitrite. Science 235: 473–476

Hibbs JB Jr, Vavrin Z, Taintor RR (1987b) L-arginine is required for expression of the activated macrophage effector mechanism causing selective metabolic inhibition in target cells. J Immunol 138: 550–565

Hibbs JB Jr, Taintor RR, Vavrin Z, Rachlin EM (1988) Nitric oxide: a cytotoxic activated macrophage effector molecule. Biochem Biophys Res Commun 157: 87–94 [Erratum Biochem Biophys Res Commun (1989) 158: 624]

Hibbs JB Jr, Taintor RR, Vavrin Z, Granger DL, Drapier J-C, Amber IJ, Lancaster JR Jr (1990) Synthesis of nitric oxide from a terminal guanidino nitrogen atom of L-arginine: a molecular mechanism regulating cellular proliferation that targets intracellular iron. In: Nitric oxide from L-arginine: a bioregulatory system. Elsevier, New York, pp 189–223

Hibbs JB Jr, Bastian NR, Taintor RR, Vavrin Z, Granger DL (1992a) Activity of the inducible high output nitric oxide synthase during immunological rejection of allogeneic P815 mastocytoma cells in Swiss-Webster mice. In: Moncada S, Marletta MA, Hibbs JB Jr, Higgs EA (eds) The biology of nitric oxide, vol 2. Enzymology, biochemistry and immunology. Portland Press, London Chapel Hill, pp 237–240

Hibbs JB Jr, Granger DL, Krahenbuhl JL, Adams LB (1992b) Synthesis of nitric oxide from L-arginine: a cytotoxic inducible pathway with antimicrobial activity. In: van Furth R (ed) Mononuclear phagocytes: biology of monocytes and macrophages. Kluwer Academic Publishers, Dordrecht Boston London, pp 279–292

Iyengar R, Stuehr DJ, Marletta MA (1987) Macrophage synthesis of nitrite, nitrate, and N-nitrosamines: precursors and role of the respiratory burst. Proc Natl Acad Sci USA 84 :6369–6373

Kilbourn RG, Belloni P (1990) Endothelial cell production of nitrogen oxides in response to interferon-γ in combination with tumor necrosis factor, interleukin-1, or endotoxin. J Natl Can Inst 82: 772–776

Kosaka H, Kazuhiko I, Kiyohiro I, Itiro T (1979) Stoichiometry of the reaction of exyhemoglobin with nitrite. Biochim Biophys Acta 581: 184–188

Kwon NS, Stuehr DJ, Nathan CF (1991) Inhibition of tumor cell ribonucleotide reductase by macrophage-derived nitric oxide. J Exp Med 174: 761–768

Lancaster JR Jr, Hibbs JB Jr (1990) EPR demonstration of iron-nitrosyl complex formation by cytotoxic activated macrophages. Proc Natl Acad Sci USA 87: 1223–1227

Lancaster JR Jr, Langrehr JM, Bergonia HA, Murase N, Starzl TE, Simmons RL, Hoffman RA (1992) Detection of iron-nitrosyl complexes by electron paramagnetic resonance spectroscopy during rejection of vascularized allograft in the rat. In: Moncada S, Marletta MA, Hibbs JB Jr, Higgs EA (eds) The biology of nitric oxide, vol 2. Enzymology, biochemistry and immunology. Portland Press, London Chapel Hill, pp 216–219

Leaf CD, Wishnok JS, Tannenbaum SR (1990) In: Moncada S, Higgs EA (eds) Nitric oxide form L-arginine: a bioregulatory system. Elsevier, Amsterdam, pp 291–299

Lepoivre M, Chenais B, Yapo A, Lemaire G, Thelander L, Tenu J-P (1990) Alterations of ribonucleotide reductase activity following induction of the nitrite-generating pathway in adenocarcinoma cells. J Biol Chem 265: 14143–14149

Liew FY, Millott S, Parkinson C, Palmer RMJ, Moncada S (1990) Macrophage killing of *Leishmania parasite* in vivo is mediated by nitric oxide from L-arginine. J Immunol 144 : 4794–4797

Mackaness GB (1971) Resistance to intracellular infection. J Infect Dis 123: 439–445

Marletta MA, Yoon PS, Iyengar R, Leaf CD, Wishnock JS (1988) Macrophage oxidation of L–arginine to nitrite and nitrate: nitric oxide is an intermediate. Biochemistry 27: 8706–8711

Mayer J, Woods M, Vavrin Z, Hibbs JB Jr (1993) Gamma interferon-induced nitric oxide production reduces *Chlamydia trachomatis* infectivity in McCoy cells. Infect Immun 61: 491–497

Nathan CF, Hibbs JB Jr (1991) Role of nitric oxide synthesis in macrophage antimicrobial activity. Curr Opin Immunol 3: 65–70

Nussler A, Drapier J-C, Renia L, Pied S, Miltgen F, Gentilini M, Maxier D (1991) L-arginine-dependent destruction of intrahepatic malaria parasites in response to tumor necrosis factor and/or interleukin 6 stimulation. Eur J Immunol 21: 227–230

Palmer RMJ, Ashton DS, Moncada S (1988) Vascular endothelial cells synthesize nitric oxide from L-arginine. Nature 153: 1251–1256

Pellat C, Henry Y, Drapier J-C (1990) IFN-γ activated macrophage: detection by electron paramagnetic resonance of complexes between L-arginine-derived nitric oxide and non-heme iron proteins. Biochem Biophys Res Commun 166: 119–125

Prütz W, Mönig H, Butler J, Land E (1985) Reactions of nitrogen dioxide in aqueous model systems: oxidation of tyrosine units in peptides and proteins. Arch Biochem Biophys 243: 125–134

Stamler JS, Singel DJ, Loscalzo (1992) Biochemistry of nitric oxide and its redox-activated forms. Science 258: 1898–1902

Stuehr DJ, Nathan CF (1989) A macrophage product responsible for cytostasis and respiratory inhibition in tumor target cells. J Exp Med 169: 1543–1555

Weinberg JB, Chapman HA Jr, Hibbs JB Jr (1978) Characterization of the effects of endotoxin on macrophage tumor cell killing. J Immunol 121: 72–80

Wink DA, Kasprzak KS, Maragos CM, Elespuru RK, Misra J, Dunams TM, Cebula TA, Koch WH, Andrews AW, Allen JS, Keefer LK (1992) DNA deaminating ability and genotoxicity of nitric oxide and its progenitors. Science 254: 1001–1003

Correspondence: Dr. J.B. Hibbs, Jr., Division of Infectious Diseases, School of Medicine, The University of Utah, Room 4B 322, 50 North Medical Drive, Salt Lake City, UT 84132, U.S.A.

Lazaroids: potent inhibitors of iron-dependent lipid peroxidation for neurodegenerative disorders

E. D. Hall and J. M. McCall

CNS Diseases Research, The Upjohn Company, Kalamazoo, Michigan, U.S.A.

Summary

A substantial body of information supports the occurrence and pathophysiological importance of oxygen radical-mediated lipid peroxidation in acute cerebral damage secondary to traumatic or ischemic injury. Moreover, peroxidative mechanisms have been implicated in the chronic neurodegenerative disorders, including Alzheimer's and Parkinson's diseases. Consequently, there has been interest in the identification of pharmacological agents with potent ability to interrupt oxygen radical formation or cell membrane lipid peroxidative mechanisms. In this regard, our laboratories have developed a novel series of potent lipid peroxidation inhibitors, known as the 21-aminosteroids or "lazaroids". One of these compounds, U-74006F or tirilazad mesylate, has shown efficacy in models of brain injury and focal cerebral ischemia. U-74006F has further been observed to antagonize excitotoxic neuronal damage in vitro and in vivo. Additionally, the compound has been found to attenuate the increased lipid peroxidation observed in Alzheimer's brain tissue, and to retard anterograde degeneration of motor nerve fibers. Another series of antioxidants, the 2-methylaminochromans, typified by the compound U-78517F, has been discovered which are even more potent and effective inhibitors of lipid peroxidation than the 21-aminosteroids.

Introduction

The importance of oxygen free radical-induced lipid peroxidation in the acute pathophysiology of central nervous system (CNS) injury and ischemia (i.e. stroke) has been fairly well established (Braughler and Hall, 1989; Hall and Braughler, 1989). Thus, efforts have been directed toward the discovery of effective lipid antioxidant compounds that can retard post-traumatic and post-ischemic neurodegeneration. The 21-

aminosteroids, or "lazaroids", are a novel series of compounds being developed for the acute treatment of traumatic or ischemic CNS injury, which have been specifically designed to localize within cell membranes and inhibit lipid peroxidation reactions. Moreover, lipid peroxidation may play a role in chronic neurodegenerative disorders including Parkinson's (Riederer et al., 1989; Dexter et al., 1989; Fahn and Cohen, 1992) and Alzheimer's (Subbarao et al., 1990; Andorn et al., 1990) diseases. Thus, novel lipid antioxidants may find utility in these disorders as well.

The 21-aminosteroids are the products of an effort to develop non-glucocorticoid steroids that duplicate the cerebroprotective pharmacology of synthetic glucocorticoid steroids. Extensive studies with the glucocorticoid steroid methylprednisolone (see Fig. 1) had indicated that large intravenous doses (30 mg/kg) could ameliorate many of the pathophysiological consequences of traumatic or ischemic injury in the CNS, and promote the functional recovery of experimentally-injured animals by inhibiting post-traumatic lipid peroxidation (Hall, 1992). The definition of this high-dose, non-glucocorticoid antioxidant action led to the synthesis of a number of non-glucocorticoid steroid analogs of methylprednisolone (e.g. U-72099E; Fig. 1) which also weakly inhibited lipid peroxidation in high concentrations, and at high doses were active in models of experimental CNS trauma (Hall et al., 1987). These non-glucocorticoid steroids became springboards for compounds that would be even more potent and effective inhibitors of lipid peroxidation, with greater activity in experimental models of CNS trauma and ischemia.

The first compound in the 21-aminosteroid series was synthesized in 1985. U-74006F (Fig. 1; generic name is tirilazad mesylate; trade name is Freedox®), has been selected for clinical development for the acute treatment of traumatic brain and spinal injury, subarachnoid hemorrhage and stroke, and is currently involved in Phase III clinical trials in each of those disorders. In the present discussion, a main focus will be on the known antioxidant mechanisms of U-74006F in particular, and the compound's effects in experimental models of brain injury and focal and global brain ischemia. Additionally, the ability of the compounds to antagonize excitotoxic neuronal injury and to reduce the increased lipid peroxidation in Alzheimer's brain tissue will be discussed. Moreover, results in a model of motor nerve degeneration will be reviewed. Another 21-aminosteroid, U-74500A (Fig. 1), is actually a more potent inhibitor of iron-catalyzed lipid peroxidation than U-74006F, but it has not been chosen for development due to chemical instability and rapid elimination in vivo. In addition, brief mention will be made of more recently discovered antioxidants, the 2-methylaminochromans, in which the steroid moiety of U-74006F has been replaced by the more potent antioxidant chromanol antioxidant structure of vitamin E (i.e. alpha tocopherol).

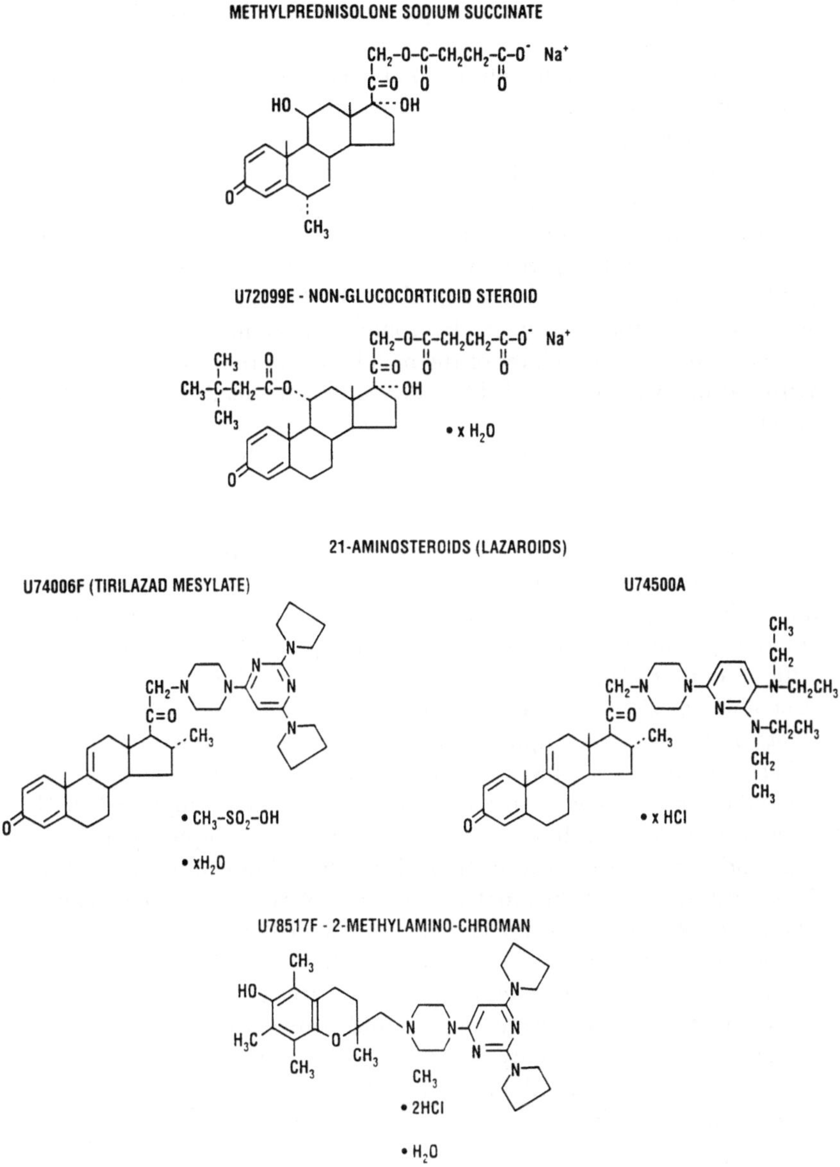

Fig. 1. Chemical structures of the glucocorticoid methylprednisolone, the non-glucocorticoid steroid U-72099E, the 21-aminosteroids U-74006F and U-74500A, and the 2-methyl-aminochroman U-78517F

Mechanisms of lipid peroxidation inhibition

U-74006F is a very lipophilic compound (log of calculated octanol/water partition coefficient = 8) that distributes preferentially to the lipid bilayer of cells. It appears that the compound exerts its effects through cooperative effects: a radical scavenging action (i.e. chemical antioxidant effect) and a physicochemical interaction with the cell membrane that serves to decrease membrane fluidity (i.e. membrane stabilization). Cytoprotection is governed partly by intrinsic reactivity toward free radicals, partly by location and orientation of U-74006F within the membrane, and partly by the ability of the compound to modify the physical properties of the lipid bilayer of the membrane. Activities in different in vitro and in vivo models probably reflect different balances between these actions.

The 21-aminosteroids are potent inhibitors of lipid peroxidation in vitro. Using rat brain homogenates or purified rat brain synaptosomes as the lipid source, U-74006F and U-74500A potently inhibit iron-dependent lipid peroxidation, with an efficacy greatly surpassing that of the glucocorticoid steroid methylprednisolone. In a model that uses synaptic membranes prepared from rat brain as a lipid source and 200 μM ferrous chloride to initiate and catalyze the lipid peroxidation reactions, U-74006F inhibited lipid peroxidation with IC_{50}s ranging from 10–60 μM (Braughler et al., 1987). U-74006F also protects isolated liver microsomes from oxidative injury that is initiated by ferrous ammonium sulfate. The IC_{50} is 3.8 μM when the U-74006F is added in ethanol to a suspension of microsomes in Krebs buffer. Very interestingly, when U-74006F is added as a lipid emulsion (triglyceride, phosphatidylcholine, drug and water) rather than in ethanol, the IC_{50} drops to below 0.1 μM. This illustrates one of the problems in testing very lipophilic compounds like U-74006F. When such compounds are added in organic solution to physiologic buffers, they microprecipitate. Emulsion delivery is often a superior delivery technique for compounds of this class.

Due to the fact that many of the lipid peroxidation models involve initiation of oxidative injury by iron, the 21-aminosteroids have been mistakenly described as inhibitors that affect exclusively iron-dependent lipid peroxidation. However, we have also studied iron-free systems and shown lipid antioxidant effects of U-74006F and U-74500A. In addition, U-74006F has been shown effective in a model of lipid peroxidation that involved rat liver microsomes with initiation by cumene hydroperoxide. Free iron was removed with a chelation column, thus insuring that the system was truly iron-free (Bryan et al., 1990). U-74006F has further been demonstrated to inhibit diquat-induced lipid peroxidation in liver microsomes (Wolfgang et al., 1991).

U-74006F and U-74500A have been reported to scavenge lipid peroxyl and phenoxy radicals in a methanol solution of linoleic acid in the presence of 2,2'-azobis(2,4-dimethylvaleronitrile, AMVN) which induces peroxidation of the polyunsaturated linoleic acid, although both U-74006F and U-74500A possess slower rate constants in this environment than the prototypical peroxyl radical scavenger vitamin E. However, both compounds act to slow the oxidation of vitamin E during linoleic acid peroxidation and potentiate vitamin E's antioxidant efficacy (Braughler and Pregenzer, 1989).

Recently, we further studied the lipid radical scavenging properties of the 21-aminosteroids in three different models (Linseman et al., 1993). Model 1 involves a homogeneous methanolic solution of linoleic acid as the peroxidizable lipid substrate in methanol, with AMVN as the free radical initiator as previously described by Braughler and Pregenzer (1989). Model 2 involves multilamellar vesicles of dilinoleyllecithin with 2,2'-azobis(2-amidinoaminopropane), ABAP, as the water-soluble initiator. In model 2, U-74006F or U-74500A were incorporated in the multilamellar vesicle as it was prepared. Both initiators are thermally activated and produce lipid-free radicals at a constant and readily reproducible rate, thereby creating a steady-state kinetic system. Hydroperoxide formation was measured by HPLC in Model 1 and by a xylenol orange color test for lipid hydroperoxides in Model 2; hydroperoxide LOOH formation was found to be linear for the time periods measured. The rates of hydroperoxide formation were proportional to the square root of the concentration of initiator and to the concentration of substrate. When U-74500A was added as the inhibitor, a transient decrease in the hydroperoxide production was observed and, during the same time period, the compound was degraded in a first-order manner. Thus, in the homogeneous system, the inhibitor acts by scavenging lipid radicals and its reactivity is about 30 times larger than that of linoleic acid. In Model 3, rat liver microsomes were treated with ferrous ammonium sulfate. This initiates an iron-mediated lipid peroxidation that is empirically described by the measure of malonyldialdehyde (MDA) that is formed. U-74006F was effective in all of these models. However, it was most effective when it is in the ordered environment of the lipid vesicle (Model 2) or the microsome (Model 3). U-74500A is actually a better antioxidant than U-74006F. It has a lower oxidation potential and this makes it a superior in vitro lipid peroxidation inhibitor.

In addition to scavenging of lipid peroxyl radicals, U-74006F also reacts with reactive oxygen species such as hydroxyl radicals generated during in vitro Fenton reactions (i.e. $Fe^{++} + H_2O_2 \rightarrow Fe^{++} + OH^- + \cdot OH$) (Althaus et al., 1991). In vivo studies employing the salicylate trapping method for measurement of hydroxyl radical have demonstrated that U-

74006F administration decreases brain hydroxyl radical levels in models of concussive head injury in mice (Hall et al., 1992, 1993) and global cerebral ischemia/reperfusion injury in gerbils (Althaus et al., 1991). This may be due to either direct scavenging of hydroxyl radical or a decrease in its injury-induced formation. This work is further detailed below.

The 21-aminosteroids have also been shown to inhibit lipid peroxidation in whole cells. For example, U-74500A inhibits copper-induced red cell lipid peroxidation. The compound is effective at concentrations as low as 1 μM. At 1 μM, it significantly reduces copper-induced and H_2O_2-induced erythrocyte lipid peroxidation by 76.5% and 27.6%, respectively. The inhibition of erythrocyte lipid peroxidation was accompanied by an inhibition of hemolysis (Fernandes, 1992). U-74006F (5 μM) has been shown to protect murine neocortical cell cultures that were exposed to 50 μM ferric iron and 50 μM ferrous iron for 24 hrs from degeneration (Monyer et al., 1990). The compound has also been reported to protect cultured murine spinal neurons from damage by 200 μM ferrous iron (Hall et al., 1991).

U-74006F is also effective in an in vitro model for predicting a compound's ability to prevent cell damage during periods of energy failure. Iodoacetic acid (IAA) was administered to the cultured human astroglial cells (UC-11MG) at a concentration of 50 μM for 4 hrs. This agent shuts down glycolysis and leads to subsequent irreversible breakdown of cellular membranes and ultimately to cell death. During the first hours after addition, IAA rapidly depleted cellular levels of ATP and decreased active uptake of tritiated aminoisobutyric acid. Subsequent irreversible cellular injuries were characterized by the release of large amounts of free arachidonic acid into the extracellular medium, massive calcium influx, and leakage of cytoplasmic contents (51Cr release). The appearance of 15-hydroxy eicosatetraenoic acid in membrane phospholipids and loss of cellular thiol groups indicated the cell constituents were being assaulted by oxidative species. These manifestations of IAA-induced cell damage were inhibited by U-74006F. IM-induced release of tritiated arachidonic acid was inhibited with an IC_{50} = 6 μM. U-74006F was effective even when it was administered up to 1 hr after the onset of the metabolic insult (Sun et al., 1992). In other work, U-74006F was also shown to decrease the release of arachidonic acid from cultured AtT-20 pituitary tumor cells triggered by exposure to either IAA or ferrous iron (Braughler et al., 1988).

The 21-aminosteroids U-74006F and U-74500A also have potent stabilizing effects on cell membranes. As noted above, the compounds have a high affinity for the lipid bilayer because of their lipophilicity. Based on it's membrane interaction, U-74006F has been shown to exert physicochemical effects on endothelial cell membranes. Bovine brain microvessel endothelial cells (BMECs) were labeled with diphenyl-hexatriene (DPH) fluorophores. Interactions with cell membranes were characterized with fluorescence anisotropy and fluorescence lifetimes. U-74500A and U-74006F preferentially altered the fluorescence anisotropy and lifetime parameters of the fluorescent DPH probe that distributed into the membranes throughout the BMECs. Little or no effect of the compounds were observed on the fluorescence parameters of the probe (TMA-DPH) that localized on the surface of BMEC plasma membranes. By contrast, cholesterol, used as a positive control, substantially altered the fluorescence parameters of BMECs labeled with either surface or membrane core probes. These experiments suggest that these 21-amino-steroids induce changes in the molecular packing order in membrane hydrophobic domains throughout the BMEC (Audus et al., 1991).

Other research has further defined the physicochemical effects of the 21-aminosteroids on membranes. U-74006F and vitamin E were studied in bilayer lipid membranes with time-resolved fluorescence depolariza-tion and angle-resolved fluorescence depolarization techniques, and by electron paramagnetic resonance-utilizing probe molecules (Van Ginkel et al., 1992). Lipid peroxidation products (oxidized fatty acids) strongly disorder the unsaturated lipid membranes they inhabit, but they do not affect the lipid dynamics. This is compatible with a model where the lipid hydroperoxy or hydroxy moieties reside closer to the polar head group region of the membrane lipids. U-74006F also has a disordering effect in the lipid systems, although the effects on dynamics vary de-pending on the surrounding lipids. Generally, U-74006F decreased dynamics (increased head group order). The decrease for U-74006F is consistent with the observations from additional work by Hinzmann et al. (1992) who have found that the 21-aminosteroids are incorporated into the lipid bilayer where they occupy strictly-defined positions and orientations. Figure 2 shows our hypothesis regarding the membrane interactions of U-74006F. We hypothesize that U-74006F resides in the cell membrane, and that the piperazine nitrogen, which is largely proto-nated (i.e. positively charged) at physiologic pH should orient with the acidic phospholipid head groups of the membrane bilayer by ionic interaction with the negatively-charged, phosphate-containing head groups. The steroid moiety, on the other hand, should localize within the hydrophobic core of the membrane. The pyrimidine amine of the molecule should help compress membrane phospholipid head groups. Indeed, head group viscosity in a lipid monolayer increases significantly with as little as 1.0 mole % (relative to lipid) of U-74006F (F. Kezdy et al., The Upjohn Co., personal communication). In addition to the com-pound's chemical antioxidant properties described above, this "mem-brane stabilizing" action may help to inhibit the propagation of lipid

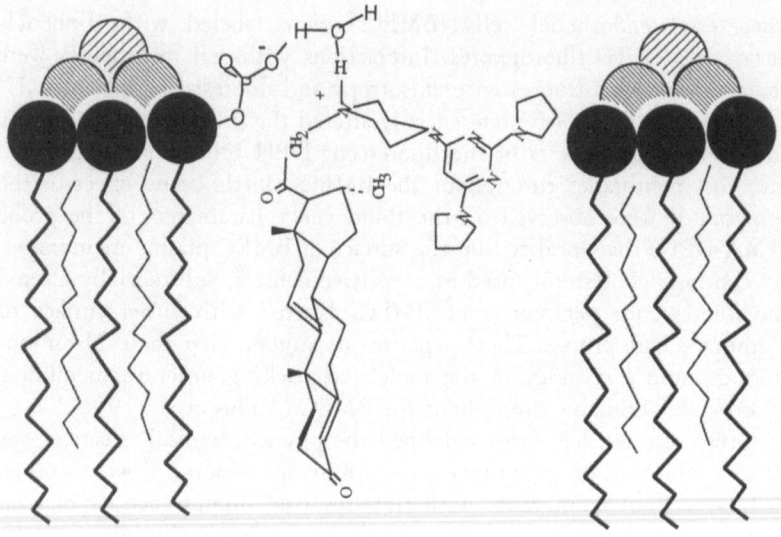

Fig. 2. Membrane interactions of the 21-aminosteroid U-74006F

peroxidation by restricting the movement of lipid peroxyl and alkoxyl radicals within the membrane.

Concerning other possible cerebroprotective mechanisms of action, neither U-74500A nor U-74006F produce hypothermic or CNS depressant effects (Hall et al., unpublished results). Moreover, they do not directly antagonize excitatory amino acid (e.g. NMDA) receptor activation in vitro (Monyer et al., 1990). Finally, they do not exhibit any significant competition for cholinergic, adrenergic, serotonergic, dopaminergic, opiate, or benzodiazepine receptors with standard ligands (Hall et al., unpublished results). Thus, the only demonstrated cerebroprotective mechanism of the 21-aminosteroids concerns their ability to block oxygen radical-induced lipid peroxidation, apparently via a combination of chemical antioxidant (i.e. radical scavenging) and membrane-stabilizing effects.

Effects in models of brain injury

Initial studies of the efficacy of U-74006F in acute head injury have been carried out to determine the ability of the compound to improve early neurological recovery and survival of head-injured mice (Hall et al., 1988b). Administration of a single i.v. dose of U-74006F produced a significant improvement in the 1-hr, post-injury neurological status (grip test score) over a broad range (0.003–30 mg/kg). A 1 mg/kg i.v. dose given within 5 mins and again at 1.5 hrs after a severe injury, in addition to improving early recovery, also increased the 1-week survival to 78.6% compared to 27.3% in vehicle-treated mice (p < 0.02). A similar reduction in 48-hr post-traumatic mortality by U-74006F has been reported in rats subjected to moderately severe fluid percussion head injury (McIntosh et al., 1992).

Recent data obtained from the mouse head injury model suggests that the compound may also reduce post-traumatic brain hydroxyl radical concentrations (Hall et al., 1992,1993). Administration of salicylate, a hydroxyl radical trap, leads to the post-traumatic formation of dihydroxy-benzoic acid (DHBA) in brain tissue when it reacts with hydroxyl radical. Severe concussive head injury results in an increase in brain levels of salicylate-derived DHBA measured at 30 mins post-injury. However, a 3 mg/kg i.v. dose of U-74006F produces a decrease in DHBA formation, implying either an attenuated formation or a chemical scavenging of hydroxyl radicals (Hall et al., 1992). While the former cannot be ruled out, the latter would seem more likely in view of in vitro studies showing that U-74006F can indeed react with hydroxyl radical as described earlier in this paper. In less severe injuries, a 1 mg/kg i.v. dose of the compound has been shown to effectively blunt the post-traumatic increase in hydroxyl radical (DHBA) levels (Hall et al., 1993).

Coincident with the reduction in brain hydroxyl radical levels, U-74006F (3 mg/kg i.v. within 5 mins post-injury) also acts to reduce post-traumatic opening of the blood-brain barrier (Hall et al., 1992). Severe concussive head injury in mice rapidly increases the uptake of ^{14}C-albumin into brain, implying an increased permeability to protein. Acute U-74006F treatment, however, restores protein permeability to normal, non-injured levels beginning with the 30-min post-injury measurement time. This effect of a single i.v. dose persists at least to 1 hr post-injury. The increased permeability of the blood-brain barrier is also apparent in an increased brain uptake of the highly protein-bound U-74006F in head injured mice compared to normal mice. The effect of U-74006F to reverse this leaky barrier may also be apparent in the fact that by 30 mins after dosing, the brain levels of U-74006F are no longer different from those observed in normal uninjured brains. This effect of U-74006F to close the barrier may be related to the attenuation of hydroxyl radical levels or an antagonism of the effects of free radicals on the barrier endothelium. Indeed, free radicals are known to increase barrier permeability (Greenwood, 1991). Consistent with the present data showing a reduction in post-traumatic blood-brain barrier opening which leads to vasogenic brain edema, U-74006F has been shown to attenuate post-traumatic brain edema in a rat model of fluid percussion head injury (McIntosh et al., 1992).

Additional experiments have been conducted in severely head-injured cats to assess the effects of U-74006F on brain energy metabolites (Dimlich et al., 1990). A 1 mg/kg i.v. dose administered at 30 mins post-injury, plus a second 0.5 mg/kg dose 2 hrs later, resulted in an improved metabolic profile within the injured hemisphere measured at 4 hrs. Most notably, U-74006F significantly reduced post-traumatic lactic acid accumulation in both the cerebral cortex and subcortical white matter. This biochemical effect suggests either an improved maintenance of cerebral blood flow or an improvement in mitochondrial function in the injured brain.

Effects in models of focal cerebral ischemia

Protective effects of the 21-aminosteroids, and U-74006F in particular, have been obtained in experimental models of focal cerebral ischemia. First of all, in a model of temporary hemispheric cerebral ischemia produced in the Mongolian gerbil by unilateral occlusion of a carotid artery, U-74006F has been examined for its ability to promote survival and reduce neuronal necrosis (Hall et al., 1988a). Gerbils were pretreated with either vehicle or U-74006F i.p. 10 mins before, and again immediately after a 3-hr temporary occlusion of the right carotid artery. U-74006F improved both 24- and 48-hr survival compared with vehicle-treated animals. Histological examination of the brains of vehicle-treated animals at 24 hrs revealed marked neuronal cell loss (as much as 90% below control) in the hippocampus and lateral cerebral cortex. In contrast, the neuronal densities in the ischemic hemisphere of gerbils treated with U-74006F showed a dose-related and statistically-significant preservation of neurons in both brain regions.

Mechanistic studies in the same model have demonstrated that neuroprotective dosing with U-74006F lessens post-reperfusion brain lipid peroxidation as evidenced by an attenuation of brain vitamin E depletion (Hall et al., 1991). A similar 2-hr, post-ischemic preservation of ascorbic acid levels has also been observed with U-74006F (Sato and Hall, 1992). The fact that this in vivo lipid antioxidant effect results in neuronal membrane protection has also been demonstrated by a U74006F-induced improvement in post-ischemic recovery of cortical extracellular calcium levels. In essence, the compound acts to preserve cellular processes (Ca^{++} ATPase; Na^+-Ca^{++} exchange mechanism) responsible for the reversal of the ischemia-triggered intracellular calcium accumulation.

U-74006F has also been examined in the cat following a 1-hr temporary occlusion of the middle cerebral artery (MCA) (Silvia et al., 1987). Beginning 15 mins after occlusion release, U-74006F was administered as a multiple bolus regimen that was continued for 12 hrs. At 1 week, using classical histological and quantitative 2-deoxyglucose autoradiographic techniques, the area of cerebral infarction was assessed. Postischemic treatment with U-74006F significantly reduced the infarction volume compared to that in vehicle-treated cats. However, in another cat study employing a longer period of temporary MCA occlusion (3 hrs), no effect of the compound was observed (Gelb et al., 1990).

The compound has further been shown to reduce infarct size in a rat (Wistar or spontaneously hypertensive) model of temporary (2-hr) MCA occlusion (Xue et al., 1992). U-74500A has also been reported to reduce infarct size in a rat temporary MCA occlusion model (Panetta et al., 1990).

U-74006F has also been found to be effective in reducing ischemic brain damage in rat models of permanent focal ischemia produced by sustained unilateral occlusion of a middle cerebral artery, at least in some strains of rats. In two different studies with Sprague-Dawley rats, U-74006F has been reported to reduce peri-infarct edema (Young et al., 1988; Lythgoe et al., 1990). In another investigation, U-74006F was shown to reduce the size of the infarct in Sprague-Dawley rats (Beck and Bielenburg, 1991). In contrast, U-74006F, in doses that effectively reduce infarct size with temporary (2-hr) MCA occlusion, is ineffective within the context of permanent occlusion in the spontaneously hypertensive rats (Xue et al., 1992).

Effects on lipid peroxidation in Alzheimer's brain tissue

In addition to a role of oxygen radical-mediated lipid peroxidative membrane damage in acute brain injury and ischemia, there is also accumulating evidence of increased oxidative stress (i.e. increased free radical production) and increased susceptibility to lipid peroxidation in Alzheimer's brains. Recent work has shown that there is a higher baseline content of thiobarbituric acid-reactive lipid peroxidation products in cerebral cortical tissue from Alzheimer's brains in comparison to age-matched, non-Alzheimer's brains (Subbarao et al., 1990; Andorn et al., 1990). In addition, in vitro induction of lipid peroxidation by iron is more intense in Alzheimer's cortical samples (Subbarao et al., 1990). These findings suggest the possible utility of lipid peroxidation inhibitors in the treatment of the disease. In this regard, it is noteworthy that U-74500A has been shown to effectively inhibit iron-induced lipid peroxidation in Alzheimer's brain samples, although its potency is less than that observed for inhibition of peroxidation in normal brain samples. The IC_{50} in normal brain tissue was 2.5 µM versus 10 µM for Alzheimer's brain samples (Subbarao et al., 1990). Nevertheless, its efficacy against lipid peroxidation in Alzheimer's brain suggests potential utility of lipid antioxidant therapy as a means to slow disease progression.

Effects on excitotoxic neuronal damage

The effects of the 21-aminosteroids have been examined against the cytotoxic effects of NMDA in cultured mouse cerebral cortical neurons (Monyer et al., 1990). U-74500A, either given before or after NMDA exposure, significantly attenuates neuronal damage by NMDA, while having no effects on NMDA-induced membrane currents. Thus, the mechanism of this protection is indirect. Additionally, both U-74500A and U-74006F reduced neuronal damage produced by peroxidative (i.e. iron-induced), hypoglycemic or hypoxic insult to the cortical cultures. Interestingly, the competitive NMDA antagonist dextromethorphan, although not an antioxidant, also decreased the damaging effects of iron. This has led to the suggestion that excitotoxic and lipid peroxidative neuronal injury mechanisms are linked. From the standpoint of relevance to Alzheimer's disease, it is intriguing to note that amyloid beta protein has been reported to exacerbate excitotoxic damage to cortical neurons (Koh et al., 1990). In addition, vitamin E has been shown to attenuate the in vitro neurotoxicity of amyloid beta protein (Schubert et al., 1992). Taken together, these studies suggest that amyloid, excitotoxicity (i.e. glutamate-induced) and iron-catalyzed, oxygen radical-induced lipid peroxidation may be interactive neurodegenerative mechanisms. In any case, it may be that lipid antioxidant therapy may be capable of inter-rupting both excitotoxic and lipid peroxidative degeneration relative to Alzheimer's pathogenesis.

Recent studies have clearly demonstrated an excitotoxic-free radical link in vivo. Boisvert (1992), using the salicylate trapping method, has shown that glutamate infusion into rat striatum via a microdialysis cannula triggers increased formation of 2,5-DHBA, a product of sali-cylate and hydroxyl radical. In other words, glutamate action triggers increased hydroxyl radical levels. The glutamate infusion also produced significant neuronal damage by 24 hrs post-infusion. In contrast, treat-ment with U-74006F (3 mg/kg i.v.) at 10 min before initiation of the glutamate infusion, significantly blunted the glutamate-induced rise in 2,5-DHBA and reduced the 24-hr neuronal necrosis. These results, together with the in vitro work of Monyer et al. (1990) strongly support the concept of a unified excitotoxic-free radical-lipid peroxidative neurodegenerative mechanism.

Effects on motor nerve degeneration

The 21-aminosteroid U-74006F has also been demonstrated to affect the rate of anterograde (Wallerian) degeneration of motor nerve fibers after experimental injury. Specifically, the effects of oral pretreatment with U-74006F have been examined on the rate of functional degeneration of cat soleus motor nerve terminals after nerve (i.e. axon) section. Cats were dosed with 7.7, 13 or 30 mg/kg doses of U-74006F twice daily for 5 days, followed by unilateral sciatic nerve section on Day 5 (Hall and Yonkers, 1990). At 48 hrs after nerve section in untreated animals, the ratio of nerve-evoked (0.4 Hz) contractile tension in the degenerating nerve muscle preparation was only 52% compared to the contralateral normal preparation. In contrast, the U-74006F-treated (13 mg/kg) cats showed a ratio of 86% ($p < 0.01$ versus untreated). U-74006F also significantly improved the ability of the degenerating preparation to maintain high frequency repetitive transmission. These results strongly suggest a fundamental mechanistic role of oxygen radical-mediated lipid peroxidation in the anterograde degenerative process. Further support for this conclusion is derived from earlier studies showing that intensive dosing with the lipid antioxidant alpha tocopherol (Hall, 1987) or the glucocorticoid methylprednisolone which possesses antioxidant activity (Hall et al., 1983), also retard motor nerve degeneration in the same model. Thus, it may be that potent antioxidants may be effective in slowing neuronal degeneration in motor neuronal or neuropathic disorders.

Non-steroidal lazaroids

One of the problems with the 21-aminosteroids, vis-a-vis the prospects of their development for chronic oral treatment of neurodegenerative disorders, is the fact that their oral bioavailability is less than 10% in various species, including man. Thus, further discovery efforts have been aimed at the possibility of achieving improved oral activity while perhaps enhancing the cerebral antioxidant activity. Our approach has been to replace the steroid functionality, which possesses only weak antioxidant activity without the complex amino substitution, with a known antioxidant. Accordingly, a series of compounds has been synthesized in which the steroid of U-74006F has been replaced by the antioxidant ring structure (i.e. chromanol) of alpha-tocopherol (vitamin E). One of these compounds, U-78517F (Fig. 1), has been demonstrated to have predictably more potent in vitro lipid antioxidant and in vivo cerebroprotective activity (Hall et al., 1991). Compounds of this type may find utility in the treatment of diseases like Parkinson's, Alzheimer's and amyotrophic lateral sclerosis, with the goal of slowing lipid peroxidation-mediated neuronal degeneration.

References

Althaus JS, Williams CW, Andrus PK, Von Voigtlander PF, Hall ED (1991) In vitro and in vivo analysis of tirilazad mesylate (U-74006F) as a hydroxyl radical scavenger. Soc Neurosci Abstr 17: 164

Andorn AC, Britton RS, Bacon BR (1990) Evidence that lipid peroxidation and total iron are increased in Alzheimer's brain. Neurobiol Aging 11: 316

Audus KL, Guillot FL, Braughler JM (1991) Evidence for 21-aminosteroid association with the hydrophobic domains of brain microvessel endothelial cells. Free Rad Biol Med 11: 361–371

Beck T, Bielenberg GW (1991) The effects of the 21-aminosteroids on overt infarct size 48 hours after middle cerebral artery occlusion in the rat. Brain Res 560: 159–162

Boisvert DPJ, Schreiber C (1992) Interrelationship of excitotoxic and free radical mechanisms. In: Krieglstein J, Oberpichler-Schwenk H (eds) Pharmacology of cerebral ischemia, 1992. Wissenschaftliche Verlagsgesellschaft, Stuttgart, pp 311–320

Braughler JM, Hall ED (1989) Central nervous system trauma and stroke. I. Biochemical considerations for oxygen radical formation and lipid peroxidation. Free Rad Biol Med 6: 289–301

Braughler JM, Pregenzer JF (1989) The 21-aminosteroid inhibitors of lipid peroxidation: reactions with lipid peroxyl and phenoxyl radicals. Free Rad Biol Med 7: 125–130

Braughler JM, Pregenzer JF, Chase RL, Duncan LA, McCall JM, Jacobsen EJ (1987) Novel 21-aminosteroids as potent inhibitors of iron-dependent lipid peroxidation. J Biol Chem 262: 10438–10440

Braughler JM, Chase RL, Neff GL, Yonkers, PA, Day JS, Hall ED, Sethy VH, Lahti RA (1988) A new 21-aminosteroid antioxidant lacking glucocorticoid activity stimulates ACTH secretion and blocks arachidonic acid release from mouse pituitary tumor (AtT-20) cells. J Pharmacol Exp Ther 44: 423–427

Bryan CL, Lawrence RA, Hall ED, Jenkinson SG (1990) 21-Aminosteroids inhibit microsomal lipid peroxidation independent of iron. FASEB J 4: A630

Dexter DT, Carter CJ, Wells FR, Javoy-Agid F, Agid Y, Lees A, Jenner P, Marsden CD (1989) Basal lipid peroxidation in substantia nigra is increased in Parkinson's disease. J Neurochem 52: 381–389

Dimlich RVW, Tornheim PA, Kindel RM, Hall ED, McCall JM (1990) Effects of a 21-aminosteroid (U-74006F) on cerebral metabolites and edema after severe experimental head trauma. In: Long D (ed) Advances in neurology, vol 52. Raven Press, New York, pp 365–375

Fahn S, Cohen G (1992) The oxidant stress hypothesis in Parkinson's disease: evidence supporting it. Ann Neurol 32: 804–812

Fernandes AC, Filipe PM, Manso CF (1992) Protective effects of U-74500A against copper-induced erythrocyte and plasma lipid peroxidation. Eur J Pharmacol 220: 211–216

Gelb AW, Henderson SM, Zhang C (1990) U-74006F, a 21-aminosteroid, does not ameliorate feline focal cerbral ischemia. J Neurosurg Anesthesiol 2: 240

Greenwood J (1991) Mechanisms of blood-brain barrier breakdown. Neuroradiology 33: 95–100

Hall ED (1987) Intensive antioxidant pretreatment retards motor nerve degeneration. Brain Res 413: 175–178

Hall ED (1992) The neuroprotective pharmacology of methylprednisolone: a review. J Neurosurg 76: 13–22

Hall ED, Wolf DL (1984) Methylprednisolone preservation of motor nerve function during early degeneration. Exp Neurol 84: 715–720

Hall ED, Braughler JM (1989) Central nervous system trauma and stroke. II. Physiological and pharmacological evidence for the involvement of oxygen radicals and lipid peroxidation. Free Rad Biol Med 6: 303–313

Hall ED, Yonkers PA (1990) Preservation of motor nerve function during early degeneration by the 21-aminosteroid anti-oxidant U-74006F. Brain Res 513: 244–247

Hall ED, McCall JM, Yonkers PA, Chase RL, Braughler JM (1987) A nonglucocorticoid analog of methylprednisolone duplicates its high dose pharmacology in models of CNS trauma and neuronal membrane damage. J Pharmacol Exp Ther 242: 137–142

Hall ED, Pazara KE, Braughler JM (1988a) The 21-aminosteroid lipid peroxidation inhibitor U-74006F protects against cerebral ischemia in gerbils. Stroke 19: 997–1002

Hall ED, Yonkers PA, McCall JM, Braughler JM (1988b) Effect of the 21-aminosteroid U-74006F on experimental head injury in mice. J Neurosurg 68: 456–461

Hall ED, Braughler JM, Yonkers PA, Smith SL, Linseman KL, Means ED, Scherch HM, VonVoigtlander PF, Lahti RA, Jacobsen EJ (1991) U-78517F: a potent inhibitor of lipid peroxidation with activity in experimental brain injury and ischemia. J Pharmacol Exp Ther 258: 688–694

Hall ED, Pazara KE, Braughler JM (1991) Effect of tirilazad mesylate on postischemic lipid peroxidation and recovery of extracellular calcium in gerbils. Stroke 22: 361–366

Hall ED, Yonkers PA, Andrus PK, Cox JW, Anderson DK (1992) Biochemistry and pharmacology of lipid antioxidants in acute brain and spinal cord injury. J Neurotrauma 9 [Suppl] 2: S425–S442

Hall ED, Andrus PK, Yonkers PA (1993) Brain hydroxyl radical generation in acute experimental head injury. J Neurochem 60: 588–594

Hinzmann JS, McKenna RL, Pierson TS, Han F, Kezdy F, Epps DE (1992) Interaction of antioxidants with depth-dependent fluorescent quenchers and energy transfer probes in lipid bilayers. Chem Phys Lipids 62: 123–138

Koh J, Yang LL, Cotman CW (1990) Beta amyloid protein increases the vulnerability of cultured cortical neurons to excitotoxic damage. Brain Res 533: 315–320

Linseman KL, Lutzke BS, McCall JM, Epps DE (1993) A simple kinetic method for determining the intrinsic reactivity of lipophilic antioxidants toward free radicals. The Toxicologist 13: 337

Lythgoe DJ, Little RA, O'Shaughnessy CT, Steward MC (1990) Effect of U-74006F on oedema and infarct volumes following permanent occlusion of the middle cerebral artery in the rat. Br J Pharmacol 100: 454P

McIntosh T, Thomas M, Smith D, Smith D (1992) The novel 21-aminosteroid U-74006F attenuates cerebral edema and improves survival after brain injury in the rat. J Neurotrauma 9: 33–46

Monyer H, Hartley DM, Choi DW (1990) 21-Aminosteroids attenuate excitotoxic neuronal injury in cortical cell cultures. Neuron 5: 121–126

Panetta JA, Phillips MC, Wolski K, Clemens JA (1990) Effects of two inhibitors of lipid peroxidation on ischemic brain damage. In: Krieglstein J, Oberpichler H (eds) Pharmacology of cerebral ischemia, 1990. Wissenschaftliche Verlagsgesellschaft, Stuttgart, pp 351–356

Riederer P, Sofic E, Rausch WD, Schmidt B, Reynolds GP, Jellinger K, Youdim MBH (1989) Transition metals, ferritin, glutathione, and ascorbic acid in Parkinsonian brain. J Neurochem 52: 515–520

Sato PH, Hall ED (1992) Tirilazad mesylate protects vitamins C and E in brain ischemia–reperfusion injury. J Neurochem 58: 2263–2268

Schubert D, Kimura H, Maher P (1992) Growth factors and vitamin E modify neuronal glutamate toxicity. Proc Natl Acad Sci USA 89: 8264–8267

Silvia RC, Piercey MF, Hoffmann WE, Chase RL, Tang AH, Braughler JM (1987) U-74006F, an inhibitor of lipid peroxidation protects against lesion development following experimental stroke in the cat: histological and metabolic analysis. Soc Neurosci Abstr 13: 1499

Subbarao KV, Richardson JS, Ang LC (1990) Autopsy samples of Alzheimer's cortex show increased peroxidation in vitro. J Neurochem 55: 342–345

Sun F, Taylor BM, Fleming WE (1992) Formation of intracellular reactive oxygen metabolites during irreversible cell injury. FASEB J 7: A658

Van Ginkel G, Muller JM, Siemsen F, van't Veld AA, Korstanje W, van Zandvoort MAM, Wratten ML, Sevanian A (1992) Impact of oxidized lipids and antioxidants such as vitamin E and lazaroids on the structure and dynamics of unsaturated membranes. J Chem Soc Faraday Trans 88: 1901–1912

Wolfgang GHI, Jolly RA, Petry TW (1991) Diquat-induced oxidative damage in hepatic microsomes: effects of antioxidants. Free Rad Biol Med 10: 403–411

Xue D, Slivka A, Buchan AM (1992) Tirilazad reduces cortical infarction following transient, but not permanent focal cerebral ischemia. Stroke 23: 894–899

Young W, Wojak JC, DeCrescito V (1988) Aminosteroid lipid peroxidation inhibitor reduces ionic shifts and edema in the rat middle cerebral artery occlusion model of regional ischemia. Stroke 19: 1013–1019

Correspondence: Dr. E. D. Hall, CNS Disease Research, The Upjohn Company, Kalamazoo, MI 49001, U.S.A.

The treatment of iron overload – psychiatric implications

M. Struck[1], **P. Waldmeier**[2], and **V. Berdoukas**[3]

[1] Clinical Research and Development CNS, [2] Research Department CNS, and [3] Clinical Research and Development IT/OTA, Ciba Ltd., Basel, Switzerland

Summary

The introduction of the iron-chelator deferoxamine in the treatment of acute iron poisoning and various chronic iron overload states such as beta-thalassaemia major has dramatically improved the prognosis of patients affected. The outcome of long-term treatment, however, heavily relies on patient compliance, which is a particular problem with a chelating agent that optimally has to be given as a s.c. infusion over several hours a day. Thus the availability of a safe and orally active iron chelator would be a major achievement.

Little is known about the effects of iron and of iron chelation therapy in the field of neuropsychiatry. It has been postulated, however, that iron-chelating agents may prevent or halt degeneration of dopaminergic neurones in patients suffering from Parkinson's disease. Iron is known to have an impact on the formation of neurotoxic free radicals.

New data, both clinical and preclinical, on the orally active iron chelator CGP 37 391 indicate, that the compound has the ability to achieve a negative iron balance in man in combination with a problematic tolerability profile. In rat it penetrates the blood-brain barrier and is a potent tyrosine and tryptophan hydroxylase inhibitor, which renders the hypothesis of being of value in Parkinson's disease unlikely. In contrast orally active bidentate iron-chelators may well represent a novel approach in the treatment of schizophrenia, if compounds with excellent tolerability can be found.

Introduction

The introduction of the iron-chelator deferoxamine in the treatment of acute and chronic iron overload has dramatically improved the prognosis of the patients affected. Today acute iron poisoning, which is a relatively common threat especially to small children accidentally ingesting iron preparations, has a fatal outcome of only about 1% (Henretig and Temple, 1984) compared to around 50% in the late 1950s (Aldrich, 1958) with deferoxamine playing an important role in this improvement.

The treatment of chronic iron overload as represented by various forms of transfusion haemosiderosis, idiopathic haemochromatosis, iron overload due to liver cirrhosis or porphyria cutanea tarda and iron overload in combination with aluminum overload in patients on chronic haemodialysis has also markedly improved. Of particular interest have been patients suffering from one of the severest form of thalassaemia, i.e. beta-thalassaemia major. A first major step in the treatment of these patients prolonging their life span was the introduction of maintenance blood transfusion in the late 1950s, which at least allowed patients to reach late teens. However, beyond this stage chronic iron overload quite often lead to death mainly caused by cardiac failure or arrhythmia. Premorbidly pubertal failure, diabetes mellitus and hypothyroidism resulted from iron overload. Today as a result of the combination of transfusion regimen and iron-chelating therapy many of these patients have a near normal life expectancy (Modell and Berdoukas, 1984) with lack of compliance being an unsolved issue in prolonged treatment.

Deferoxamine is commonly well tolerated (Bentur et al., 1991; Hershko and Weatherall, 1988). However, skin reactions mainly at the injection site may occur. In some cases true hypersensitivity has been reported. Visual, auditory and peripheral neurotoxicity have to be considered rare events, mostly related to high doses of deferoxamine (Bentur et al., 1991; Marciani et al., 1991; Porter and Huehns, 1989). Concomitant use of neuroleptics such as prochlorperazine had been potentially dangerous, especially in patients with auto immune disorders, such as rheumatoid arthritis with an assumed impaired blood-brain barrier (Bentur et al., 1991). Cardiovascular problems (hypo- and hypertension) may arise with rapid i.v. infusion. Extremely high doses of deferoxamine can cause severe pulmonary problems including fatal interstitial pneumonia (Castriota Scanderbeg et al., 1990). Unusual infections, in particular Yersiniosis, have occurred with the use of deferoxamine (Boyce et al., 1985).

Due to the uncomfortable route of administration possibly resulting in poor compliance, research had been focussing on the discovery of orally active iron chelators. CGP 37 391 (L1 or CP 20) has been the first compound investigated in man on a broader basis since 1983 outside the pharmaceutical industry (Agarwal et al., 1992; Olivieri et al., 1990). Until recently adequate (long-term) toxicity studies in animals were missing (Olivieri et al., 1990).

In the field of neuropsychiatry chelating agents received particular attention in the treatment of Alzheimer's disease (Crapper McLachlan et al., 1991) and in the aetiology of Parkinson's disease, where basal lipid peroxidation turned out to be increased (Dexter et al., 1989). This finding was later related to increased concentrations of iron in the substantia nigra (Sofic et al., 1991). Melanin containing dopaminergic neurones had been recognized to be particularly susceptible to degeneration (Hirsch et al., 1988). Melanin itself has a high capacity of binding iron (Ben-Shachar et al., 1991). Thus it was postulated that it might play a role in the degeneration process via the formation of free radicals in the presence of sufficient iron. As a consequence the removal of iron from the substantia nigra with chelating agents was considered to have beneficial effects. In one study in rats, intraventricular application of deferoxamine was shown to prevent dopaminergic lesions caused by 6-hydroxy-dopamine (Ben-Shachar et al., 1992).

Material and methods

As a result of the effort to evaluate the pharmacological potential of various new orally active iron chelators longterm animal toxicity studies primarily with CGP 37 391 had been initiated. All published data on the use of this compound in man were analysed. Access to all available human data was sought. In addition, the substance was screened for possible neuropsychiatric effects in mice and rats, since the compound was likely to penetrate the blood-brain barrier and its catechol structure led to the suspicion of the compound being a COMT inhibitor.

Results

Clinical data from 104 patients worldwide suffering mostly from beta-thalassaemia major indicate that it is possible to achieve a negative iron balance with CGP 37 391. Previous findings from animal studies indicating that the mean iron excretion depends on the dose applied and that the dose response curve is steep were confirmed in humans. At 25 mg/kg/day the drug is ineffective. 50 mg/kg/day seems to be the lower limit at which the compound effects urinary iron excretion. The application of 100 mg/kg/day leads to a more than twofold increase in iron excretion compared to 50 mg/kg/day with mean iron excretion values of about 40 mg/day. The iron complex is almost exclusively excreted via the kidneys. Ferritin levels have reduced significantly in some of the trials.

100 mg/kg/day, however, seems to be the upper dose limit in terms of acceptable tolerability in man. Thus, at the moment one has to assume a very narrow therapeutic window. In the absence of any clinical trial to-date fulfilling the European and US GCP guidelines, tolerability data have to be observed with caution. Re-analysis of the data available

indicated that neutropenia is not unlikely to occur, with an incidence of
3/104 patients (approximately 3%). This was in line with long-term
toxicology findings in various animal species, where a high incidence of
infections and toxicity to proliferating tissues were found. Bone marrow
depression, thymic and testicular atrophy were seen depending on the
dose applied. 7 out of 104 patients (7%) died due to serious adverse
events presumably not related to drug treatment. 28% of the patients
developed joint pain or arthralgia, 10% suffered from joint effusions,
16% experienced GI tract disturbances, 8% showed clinically relevant
elevations of liver enzymes and 4% zinc deficiency. Regular neuropsychiatric assessments had not been performed in any of these trials. A risk-
benefit analysis carried out by international opinion leaders, however,
still favoured the further development of this compound for the treatment of a life-threatening disease, such is beta-thalassaemia major.

Since CNS effects were not observed in man except for overdose of this
compound animal studies turned out to be of particular importance.
Single dose p.o. administration in mice and rats up to 100 mg/kg did not
reveal behavioural changes in both species. Hyperhidrosis was seen in rats
at doses higher than 100 mg/kg. In much higher doses, however, marked
CNS effects could be demonstrated (Gentsch et al., 1992). In mice after
single p.o. administration of 300 and 1000 mg/kg hypoactivity, ataxia
and ptosis could be observed. Passive avoidance behaviour was significantly deteriorated. After having received 1000 mg/kg 3 out of 4 mice
died. In rats motility was decreased after 300, 600 and 1000 mg/kg were
given as single dose. Food and/or water intake was impaired at 300 and
1000 mg/kg. Body temperature was attenuated after a single p.o. dose of
600 and 1000 mg/kg. Screening for CNS effects in animals following
long-term treatment especially on low dose regimen has not yet been
completed.

Since the structure of the compound raised suspicion that it might
possess COMT-inhibiting properties, the effects of CGP 37 391 on rat
striatal and cortical monoamine metabolism were thoroughly investigated (Waldmeier et al., 1993).

As a first result it was concluded that CGP 37 391 – in contrast to
deferoxamine – penetrates the blood-brain barrier. The effects on striatal
dopamine and serotonin and their metabolite levels were impressive. At
100 mg/kg given i.p. the compound induced a rapid decrease of
dopamine and HVA levels. Two hours after administration dopamine
levels were reduced by almost 50% with a near total decrease of HVA.
Control levels for dopamine were again reached 6 hours post-treatment;
those for HVA and DOPAC only after a period of 16 hours.

Similar effects could be seen for serotonin and its metabolite 5-

hydoxyindoleacetic acid with a maximum effect of a 50 to 60 percent reduction on 5-HT after 2 hours and on 5-HIAA after 4 hours. Control levels were reached again only after 16 hours post-treatment.

The observed effects both on dopamine and serotonin metabolism were dose-dependent. The impact on main metabolites already reached statistical significance at doses as low as 1 mg/kg i.p., where both HVA and 5-HIAA were reduced by about 20 percent. DA and 5-HT showed a statistically significant reduction of about 25–30% at a dose of 30 mg/kg; the maximum effect of an approximately 50% reduction being observed at 100 mg/kg. At this dose HVA was almost not measurable any more; 5-HIAA was reduced to a lower extent, with a maximum reduction of 40–50% in the dose range of 30 to 100 mg/kg.

The centrally acting decarboxylase inhibitor NSD 1015 was used to demonstrate that the administration of CGP 37 391 leads to a complete suppression of the accumulation of the precursors DOPA and 5-HTP both in striatum and cortex when given 30 minutes or 1 hour prior to CGP 37 391.

Similar effects to the ones mentioned above could not be observed with deferoxamine.

Discussion

Screening for haemoglobin, serum iron and transferrin has been part of the workup of psychiatric disorders for years. It is known that both anaemia and iron overload may be associated with depression and psychosis (Cutler, 1991; Kaplan et al., 1991). Developmental delays in iron-deficient anaemic children have been successful treated with dietary ferrous sulphate (Idjradinata and Pollitt, 1993). However, the role of iron in various neuropsychiatric disorders still remains unclear.

Recently it has been hypothesised that treatment with iron-chelators in neurodegenerative disorders such as Parkinson's disease may lead to a reduction of free radical formation and subsequently to a reduction of basal lipid peroxidation (Ben-Shachar et al., 1991, 1992), a major cause of neuronal death.

Deferoxamine, a hexadentate iron-chelator with low lipophilicity does not enter the brain after it has been administered parenterally. Also, it is not understood how peripheral iron depletion affects iron concentration and iron distribution in the adult brain (Shoulson, 1992). Therefore the clinical potential of this compound in the treatment of Parkinson's disease seems to be limited.

In contrast, deferoxamine has shown a significant reduction in the rate of decline of daily living skills in a pilot study in dementia of the

Alzheimer type (Crapper McLachlan et al., 1991). Also, a considerable number of patients suffering from dialysis encephalopathy associated with chronic aluminum overload have improved (Ackrill et al., 1980; Ackrill and Day, 1985; Arze et al., 1981) with regular deferoxamine treatment.

So far some 14.000 patients mainly suffering from beta-thalassaemia major have been treated with deferoxamine. Visual, auditory and peripheral neurotoxicity turned out to be a problem related to long term treatment, administration of high doses and low ferritin serum levels (Bentur et al., 1991; Cohen et al., 1991). In patients being treated with deferoxamine as an anti-inflammatory agent for rheumatoid arthritis concomitant treatment with the neuroleptic prochlorperazine led to transient coma in 2 patients (Bentur et al., 1991). It was hypothesised that the compound might have penetrated into the brain due to an impaired blood-brain barrier associated with the underlying autoimmune disorder. A recently submitted publication, however, states that such an impairment may be a direct consequence of neuroleptic treatment, particularly with phenothiazines (Ben-Shachar, personal communication).

In comparison to deferoxamine only about 120 patients had been on treatment with the lipophilic bidentate iron chelator CGP 37 391. Neurotoxicity was not observed. Due to the limited experience with this compound in man, however, it remains unclear, whether neurotoxicity is linked to iron-chelation in general or reflects a deferoxamine specific problem. Similar to deferoxamine, CGP 37 391 may also exhibit a certain potential to encourage bacterial growth. Due to toxicity problems the further clinical development of CGP 37 391 was terminated by CIBA recently.

It seems reasonable to assume that hexadentate iron-chelators might have advantages over bidentate chelators if being used as indirect radical scavengers (Porter et al., 1989). In the presence of excess iron the latter group might exert a neuromelanin-like effect in mediating the formation of free hydroxyl radicals and enhancing protein fragmentation. The stability of the iron complex is another critical factor for any new iron chelator, since redistribution of iron to potentially more toxic sites in the body may cause unexpected hazards (Porter et al., 1989; Porter and Huehns, 1989).

The effects of CGP 37391 on monoamine metabolism were impressive and had been previously discussed (Waldmeier et al., 1993). The marked decrease of HVA was compatible with a COMT-inhibiting effect. However, the decrease of DOPAC and the vast reduction of DA pointed towards a more complex mode of action. The compound evidently inhibits tyrosine as well as tryptophan hydroxylase, which are iron-dependent enzymes. Similar effects of bipyridil and o-phenantroline on tyrosine hydroxylase had been reported by Taylor and co-workers some years ago (Taylor et al., 1969). Although psychotropic effects of CGP 37391 have not been observed in patients suffering from beta-thalassaemia major so far, it remains an interesting hypothesis, whether the use of non-toxic orally active iron-chelators may represent a novel approach to antipsychotic treatment.

References

Ackrill P, Day JP (1985) Desferrioxamine in the treatment of aluminum overload. Clin Nephrol 24 [Suppl] 1: S94–S97

Ackrill P, Ralston AJ, Day JP, Hodge KC (1980) Successful removal of aluminum from patient with dialysis encephalopathy. Lancet ii: 692–693

Agarwal MB, Gupte SS, Viswanathan C, Vasandani D, Ramanathan J, Neena Desai, Puniyani RR, Chhablani AT (1992) Long-term assessment of efficacy and safety of L1, an oral iron chelator, in transfusion dependant thalassaemia: Indian trial. Br J Haematol 82: 460–466

Aldrich RA (1958) Acute iron toxicity. In: Wallerstein RO, Mettier SR (eds) Iron in clinical medicine. University of California Press, Berkeley, p 93

Arze RS, Parkinson IS, Cartlidge NEF, Britton P, Ward MK (1981) Reversal of aluminum dialysis encephalopathy after desferrioxamine treatment. Lancet ii: 1116

Ben-Shachar D, Riederer P, Youdim MBH (1991) Iron-melanin interaction and lipid peroxidation: implications for Parkinson's disease. J Neurochem 57: 1609–1614

Ben-Shachar D, Eshel G, Riederer P, Youdim MBH (1992) Role of iron and iron chelation in dopaminergic-induced neurodegeneration: implications for Parkinson's disease. Ann Neurol 32 [Suppl]: 105–110

Bentur Y, McGuigan M, Koren G (1991) Deferoxamine (Desferrioxamine): new toxicities for an old drug. Drug Safety 6 1: 37–46

Boyce N, Wood C, Holdsworth S, Thomson NM, Atkins RC (1985) Life threatening sepsis complicating heavy metal chelation therapy with desferrioxamine. Aust NZ J Med 15 5: 654–655

Castriota Scanderbeg A, Izzi GC, Butturini A, Benaglia G (1990) Pulmonary syndrome and intravenous high-dose desferrioxamine. Lancet 336: 1611

Cohen A, Martin M, Mizanin J, Konkle DF, Schwartz E (1991) Vision and hearing during deferoxamine therapy. J Pediatr 117 2: 26–330

Crapper McLachlan DR, Dalton AJ, Kruck TP, Bell MY, Smith WL, Kalow W, Andrews DF (1991) Intramuscular desferrioxamine in patients with Alzheimer's disease. Lancet 337 (8753): 1304–1308

Cutler P (1991) Iron overload in psychiatric illness. Am J Psychiatry 148 1: 147–148

Dexter Dt, Carter CJ, Wells FR, Javoy-Agid F, Lees A, Jenner P, Marsden CD (1989) Basal lipid peroxidation in substantia nigra is increased in Parkinson's disease. J Neurochem 52: 381–389

Gentsch C, Vassout A, Mondadori C (1992) CNS evaluation of CGP 37 391 (L1). CIBA Internal Report

Henretig FM, Temple AR (1984) Acute iron poisoning in children. Clin Lab Med 4 3: 575–586

Hershko C, Weatherall DJ (1988) Iron-chelating therapy. CRC Crit Rev Clin Lab Sci 26 4: 303–345

Hirsch E, Graybiel Am, Agid YA (1988) Melanized dopaminergic neurones are differently susceptible to degeneration in Parkinson's disease. Nature 334: 345–348

Idjradinata P, Pollitt E (1993) Reversal of developmental delays in iron-deficient anaemic infants treated with iron. Lancet 341 (8836): 1–4

Kaplan HI, Sadock BJ (1991) Synopsis of psychiatry, behavioral sciences. In: Clinical psychiatry, 6th edn. Williams & Wilkins, Baltimore, p 211

Marciani MG, Cianiulli P, Stefani N, Stefanini F, Peroni L, Sabbadini M, Maschio M, Trua G, Papa G (1991) Toxic effects of high-dose deferoxamine treatment in patients with iron overload: an electrophysiological study of cerebral and visual function. Haematologica 76 2: 131–134

Modell B, Berdoukas V (1984) The clinical approach to thalassaemia. Grune & Stratton (Harcourt Brace Jovanovich)

Olivieri NF, Koren G, Hermann C, Bentur Y, Chung D, Klein J, St. Louis P, Freedman MH, McClelland RA, Templeton DM (1990) Comparison of oral iron chelator L1 and desferrioxamine in iron-loaded patients. Lancet 336: 1275–1279

Porter JB, Huehns ER (1989) The toxic effects of desferrioxamine. Bailliere's Clin Haematol 2 2: 459–474

Porter JB, Huehns ER, Hider RC (1989) The development of iron chelating drugs. Bailliere's Clin Haematol 2 2: 257–292

Shoulson I (1992) Neuroprotective clinical strategies for Parkinson's disease. Ann Neurol 32 [Suppl]: 143–145

Sofic E, Paulus W, Jellinger K, Riederer P, Youdim MPH (1991) Selective increase of iron in substantia nigra zona compacta of parkinsonian brains. J Neurochem 56: 978–982

Taylor RJ, Stubbs CS, Ellenbogen L (1969) Tyrosine hydroxylase inhibition in vitro and in vivo by chelating agents. Biochem Pharmacol 18: 587–594

Waldmeier P, Buchle AM, Steulet AF (1993) The orally active iron chelator, 1,2-dimethyl-3-hydroxypyridin-4-one (L1, CP 20) inhibits COMT as well as tyrosine and tryptophan hydroxylase in rat brain in vivo. Biochem Pharmacol (in press)

Correspondence: Dr. M. Struck, Clinical Research and Development CNS, K-147.P.09 Ciba Ltd., CH-4002 Basel, Switzerland.

Iron therapy: Pros and Cons

G. Stern

Department of Neurology, Middlesex and University College Hospitals School of Medicine, University College, London, United Kingdom

Summary

The clinical, neuropathological and experimental evidence for implicating iron in the genesis of certain degenerative diseases of the central nervous system is briefly considered. Even if focal accumulation by decompartmentalisation of iron is a significant event – primarily or secondarily – and an effective selective chelating agent was available, demonstration of efficacy might prove difficult.

"The fox knows many things, but the hedgehog knows one big thing". This verse which survives only in quotation (attributed to Archilochus by Berlin, but may have been a quotation from another poem by the mock-heroic Margites which later appears in Plutarch [T. S. Pattie]) has figuratively been assumed to distinguish monists from pluralists. In the present sense, it could be applied to the previous contributors at this workshop who have endeavoured to implicate iron in disorders of the central nervous system. While both "species" have clearly been energetic and persuasive, the question whether disorder of a single element – iron – plays a significant primary aetiological role in diseases of the central nervous system remains unresolved. I have been asked to consider whether a "hedgehog" therapeutic intervention is justified or practicable.

From the impressive previous vulpine presentations, the following conclusions may be drawn by an empirical clinician.

Exogenous iron

There is no evidence to suggest that an increase in exogenous extra-cranial iron – save for penetrating wounds – can pass an intact blood-brain barrier and cause increasing total brain concentrations of iron. The control mechanisms of iron homeostasis are far too effective. For example, as far as increases in endogenous iron are concerned, studies of communities with high iron dietary intake illustrate this efficiency. The inhabitants of Ethiopia have the peculiar distinction of the highest intake of iron per person per day in the world (Hofvander, 1968). The average intake is 471 mg (range 98–1418 mg). This has been attributed to the main staple cereal crop, tef (eragrostis abyssinica), with an exceedingly high iron content of 100 mg per 100 G of cereal. Despite this huge intake, autopsy studies of liver show only mild iron overload and whereas studies of bone marrow show increasing haemosiderin deposits with increasing age, such changes are not gross and indicate the extraordinary efficiency of the mechanisms controlling absorption from the gut. It is therefore perplexing that reports continue to be published which attempt to correlate exposure to heavy metals including iron in the community with prevalence of Parkinson's disease (Rybicki et al., 1993). It is implausible that a demonstrable increase in environmental exposure to iron can be causally related to prevalence of Parkinson's disease. If there is such a relationship, which is doubtful, an explanation must be forthcoming to explain how such a putative environmental metallic toxin can enter the brain, mysteriously bypassing the gut and the blood-brain barrier when there is no evidence of increased total brain iron in health or disease. Even in conditions when iron is deposited in excessive amounts in the viscera – such as thalassaemia and haemochromatosis – there are no excessive deposits of iron in the brain. While the latter may be treated by venesection, it is ironic (if the pun may be permitted by the editor) that James Parkinson who had this form of therapy at his disposal did not recommend it for the disease paralysis agitans which was subsequently given his name (he preferred the application of liniments and vesicatories).

Focal deposits of iron in the brain and diseases of the central nervous system

Earlier in this workshop Jellinger gave a scholarly and comprehensive account of the neuropathological changes in a diversity of diseases of the central nervous system, characterised by discernible focal accumulations of iron. Intracerebral ferruginous encrustations, often with calcerous encrustations, have long been recognised in routine autopsy material. Classical studies such as those of Pick (1903) fascinated neuropathologists from the very beginning and provoked controversy about the significance of ferrocalcareous concrements. Since iron specific stains such as Perl's reaction were widely available, specific and sensitive, the presence of iron was extensively studied and reported. The seminal study of Hallervorden and Spatz (1922) is well known. In a study of five sisters in a sibship with movement disorders, a rusty brown discolouration of the pallidum and zona reticulata was clearly evident and the sulphur-ammonia test of Spatz on fresh brain was strongly positive. While the identity and characteristics, both clinical and neuropathological, of the Hallervorden-Spatz disease or syndrome remains enigmatic not even the most fervent advocate has seriously proposed that the malady should be considered as a primary siderosis of the brain and that iron is the primary aetiological toxic agent.

In addition to the examples cited by Jellinger at this workshop, an exotic and dramatic illness recognised by veterinarians since 1954 (Cordy) may be mentioned. Equine nigro-pallidal encephalomalacia occurs in horses grazing for prolonged periods on Yellow Star thistle (Centaurea solstilias). After a critical period of exposure the animals develop an acute neurological illness with profound motor disorders and show large discrete areas of colliquative necrosis in the substantia nigra and globus pallidus. The precise mechanism is unknown (Mettler and Stern, 1962), but excitotoxic amino-acids may be implicated (Roy et al., 1993). At the margins of the necrotic lesions, focal depositions of iron are abundant, providing further evidence of the non-specificity of focal iron accumulation in the diseased basal ganglia.

Recently introduced sophisticated methods of visualising intracerebral iron in a number of diseases of the central nervous system such as Parkinson's disease (Drayer, 1986) and Huntington's chorea (Kozachuk, 1986) appear to be of interest only in so far as confirming what has been known by classical (and less expensive) neuropathological methods for many years, although it is certainly true that serial MRI studies might document the rate of iron accumulation say in the anterior region of the globus pallidus in Huntington's chorea, for example – but whether such dynamic studies will do any more than endorse the findings of classical neuropathological studies is far from certain; whether such sophisticated studies might lead to therapeutic advances is even less certain.

Decompartmentalisation of brain iron and its significance

While total brain iron does not increase in diseases with preservation of the blood-brain barrier, many studies – the outstanding have been reviewed earlier in this workshop – have unequivocally demonstrated the pattern of regional iron distribution in mammalian brain and that the highest iron content is present in the substantia nigra and globus pallidus. Equally convincing are the reports of increased iron content of

the substantia nigra in Parkinson's disease. There appears to be unresolved controversy about the distribution of ferritin.

The notion that iron participates directly in oxygen free radical
formation, that lipid peroxidation results from the formation of oxygen
free radicals and that the cytotoxic hydroxyl radical is dependent on the
alteration of the redox state of iron between its ferrous and ferric valences
and the presence of hydrogen peroxide is generally accepted (Sofic et al.,
1991). Whether this state of so-called "induced selective oxidative stress"
is a significant primary or even secondary factor in the mechanisms
responsible for neurodegeneration is far more speculative.

The organisers of this workshop and their colleagues have critically
examined the concept of oxidative stress due to "siderosis of the substantia nigra" and have designed experiments to try and explain how iron
could be shunted into particular portions of the brain, why the substantia
nigra with its normal high iron content should accumulate further iron,
where this excessive iron is deposited, intra- or extra-neuronally, the
relationship to melanised dopamine neurons, how iron is thus bound and
the potential of iron chelators (Ben-Shachar et al., 1991). As a clinician,
I am not competent to evaluate critically their evidence and hypotheses.
I can only compound their difficulties by speculating whether the effects
of excessive depositions of iron in certain clearly demarcated foci within
the brain may have consequences other than explained by current neurochemical concepts. Ferro-electricity is a topical and burgeoning area of
interest in geophysics and in particular with respect to perovskite oxides
(Cohen, 1992). For example it is claimed that ferro-electric instability is
not due to shell-model-like polarisability of the oxygen iron; computa-
tional studies appear to show the importance of covalency in driving
ferro-electric instability. It may be that we have only just begun to
understand the significance of alterations in iron content at a molecular
level and that prevailing notions of "intoxication" may prove to be
superficial and incomplete.

Therapeutic potential

Whatever the explanation or consequences of focal increases of iron in
the brain in certain neurodegenerative diseases, and even accepting that
such accumulation cannot be a primary aetiological-toxic event and at
best secondary or consequential to a cascade of neurochemical disequilibria, the question of possible therapeutic intervention should be considered. The fact that unfortunately we have no rational explanation for the
primary cause of sporadic non-familial neurodegenerative disorders such
as the diseases of Parkinson and Alzheimer and motor neurone disease,

should not inhibit or discourage us from attempting to correct chemical disequilibria if there is reasonable evidence that they might be implicated at some stage in the degenerative process and the release of disability. (The fact that the cause or causes of Parkinson's disease remain obscure did not prevent the evolution of levodopa therapy). However, this inspiration should not delude us about the magnitude of the quantum gap between suggestive laboratory results in rattus rattus and effective treatment in homo sapiens. Unfortunately there is no paucity of illustrations to emphasise these profound difficulties. A recent example is α-tocopherol and the prevention of deterioration in early Parkinson's disease. On the theoretical basis that disturbed oxidative mechanisms and formation of free radicals might play a significant role in the natural history of otherwise untreated Parkinson's disease and that tocopherol, a component of vitamin E that traps free radicals, might delay the onset of parkinsonian disabilities, considerable effort was applied in a multi-centred controlled clinical trial. Unfortunately no protective effect could be demonstrated (Parkinson's Study Group, 1993). This negative result might have been anticipated by the experience of a blood-stock expert. Twenty years before the onset of his disease he formed a notion that horses ran faster when given regular vitamin E supplements. Inspired by this refutable equine conclusion, thereafter he took at least 400 IU daily for twenty years. Generous vitamin E supplements over two decades did not prevent him developing Parkinson's disease; nor, subsequently, did it materially alter the history of his illness or his response to levodopa and the subsequent development of dyskinesia and fluctuations (Stern, 1987).

Fortunately hope springs eternal and others have proposed iron chelators such as certain aminosteroids which have demonstrable relative protective activity in animal models of cerebral ischaemia and which cross the blood brain barrier (Braughler et al., 1987). Even if an effective and selective agent – chelating or otherwise – became available, considerable difficulties would be encountered in designing convincing trials of clinical efficacy. The agent would need to be demonstrably safe over prolonged periods of administration, indistinguishable from placebo, reasonably cheap, and well tolerated. To unequivocally demonstrate that such an agent had a protective or ameliorating benefit on the natural history of the illness controlled trials would need to include large numbers of patients observed for long periods – perhaps decades. Even if the agent was demonstrably effective in removing iron from the brain – for example, by classical balance studies or even serial magnetic resonance imaging techniques – this in itself would be insufficient; worthwhile symptomatic relief and significant alteration in the clinical profile of those who took the active substance would need to be shown. Considerable, but not insuperable problems – nil desperandum.

References

Ben-Shachar D, Eshel G, Finberg APM, Youdim MBH (1991) The iron chelator desferrioxamine (Desferal) retards 6-hydroxy-dopamine-induced degeneration of nigrostriatal dopamine neurons. J Neurochem 56: 1441–1444

Berlin I (1953) The Hedgehog and the Fox. Weidenfeld & Nicolson

Braughler JM, Pregenzer JF Chase RL, Ducan LA, Jacobson EJ, McCall JM (1987) Novel 21-aminosteroids as potent inhibitors of iron-dependent lipid peroxidation. J Biol Chem 262: 10438–10440

Cohen RE (1992) Origin of ferro-electricity in perovskite oxides. Nature 358: 136

Cordy DR (1954) Nigro-pallidal encephelomalacia in horses associated with ingestion of Yellow Star thistle. J Neuropathol Exp Neurol 13: 330–342

Drayer BP, Olanow CW, Burger F (1986) High field strength resonance imaging in patients with Parkinson's disease. Neurology 36: 1: 309–311

Hallervorden J, Spatz H (1922) Eigenartige Erkrankung im extrapyramidalen System mit besonderer Beteiligung des Globus pallidus und der Substantia nigra. Z Ges Neurol Psychiat 79: 254–302

Hofvander NY (1968) Haematological investigations in Ethiopia with special reference to a high iron intake. Acta Med Scand [Suppl] 494: 1965–1967; Neurology 36 (1): 310

Mettler FA, Stern GM (1962) Observations on the toxic effect of Yellow Star thistle. J Neuropathol Exp Neurol 22: 164–169

Parkinson Study Group (1993) Effects of tocopherol and deprenyl on the progression of disability in early Parkinson's disease. N Engl J Med 176–182

Pattie TS. The Manuscript Collection. The British Library (personal communication)

Pick A (1903) Weiterer Beitrag zur Pathologie der Tetanie nebst einer Bemerkung zur Chemie verkalkter Hirngefäße. Neurol 262: 22: 754-756

Roy DN, Craig AM, Blythe LL, Kayton RE, Spencer PS (1993) Partial isolation of excitotoxic agents from Yellow Star thistle (Centauria solstialis). A cause of equine nigro-striatal degeneration. Neurodegeneration (in press)

Rybicki BA Johnson CJ, Yuman J, Gorell AM (1993) Parkinson's disease mortality and the industrial use of heavy metals in Michigan. Mov Disord 8: 1: 87–92

Sofic E, Paulus W, Jellinger K, Riederer P, Youdim MBH (1991) Selective increase of iron in substantia nigra and zona compacta in parkinsonian brain. J Neurochem 56: 978–982

Stern GM (1987) Vitamin E and Parkinson's disease. Lancet i: 508

Correspondence: Dr. G. Stern, Department of Neurology, The Middlesex Hospital, Mortimer Street, London W1N 8AA, United Kingdom.

Subject Index

Toshiharu Nagatsu, Hirotaro Narabayashi,
Mitsuo Yoshida (eds.)

Parkinson's Disease. From Clinical Aspects to Molecular Basis

(Key Topics in Brain Research)
1991. 52 figures. VIII, 220 pages.
Soft cover DM 98,–, öS 686,–. ISBN 3-211-82272-0

Konrad Maurer, Peter Riederer, Helmut Beckmann (eds.)

Alzheimer's Disease. Epidemiology, Neuropathology, Neurochemistry, and Clinics

(Key Topics in Brain Research)
1990. 118 figures (9 in color). XIX, 581 pages.
Soft cover DM 176,–, öS 1230,–. ISBN 3-211-82197-X

Horst Przuntek, Peter Riederer (eds.)

Early Diagnosis and Preventive Therapy in Parkinson's Disease

(Key Topics in Brain Research)
1989. 59 figures (1 in color). XIV, 442 pages.
Soft cover DM 135,–, öS 950,–. ISBN 3-211-82080-9

Prices are subject to change without notice.

Springer-Verlag Wien New York

Sachsenplatz 4–6, P.O.Box 89, A-1201 Wien · 175 Fifth Avenue, New York, NY 10010, USA
Heidelberger Platz 3, D-14197 Berlin · 37-3, Hongo 3-chome, Bunkyo-ku, Tokyo 113, Japan

K.F. Tipton, M.B.H. Youdim, C.J. Barwell,
B.A. Callingham, G.A. Lyles (eds.)

Amine Oxidases: Function and Dysfunction

Proceedings of the 5th International Amine Oxidase Workshop, Galway, Ireland, August 22-25, 1992

1994. 113 figures. Approx. 410 pages.
Soft cover DM 220,–, öS 1540,–
Reduced price for subscribers to "Journal of Neural Transmission":
Soft cover DM 198,–, öS 1386,–
ISBN 3-211-82521-5

Journal of Neural Transmission / Supplementum 41

E. Tolosa, R. Duvoisin, F. F. Cruz-Sanchez (eds.)

Advances in Progressive Supranuclear Palsy

1994. 68 figures. Approx 256 pages.
Soft cover DM 135,–, öS 945,–
ISBN 3-211-82541-X

Subscribers to "Journal of Neural Transmission" will receive this monograph as Supplementum 42 to the journal.

Prices are subject to change without notice.

Springer-Verlag Wien New York

Sachsenplatz 4–6, P.O.Box 89, A-1201 Wien · 175 Fifth Avenue, New York, NY 10010, USA
Heidelberger Platz 3, D-14197 Berlin · 37-3, Hongo 3-chome, Bunkyo-ku, Tokyo 113, Japan